Seminal Ideas and Controversies in Statistics

Statistics has developed as a field through seminal ideas and fascinating controversies. *Seminal Ideas and Controversies in Statistics* concerns a wide-ranging set of 15 important statistical topics, grouped into three general areas: philosophical approaches to statistical inference, important statistical methodology for applications, and topics on statistical design, focusing on the role of randomization. The key papers on each topic are discussed with commentaries to help explain them. The goal is to expand reader knowledge of the statistics literature and encourage a historical perspective.

Features

- Discusses a number of important ideas in the history of statistics, including the likelihood principle, Bayes vs. frequentist approaches to inference, alternative approaches to least squares regression, shrinkage estimation, hypothesis testing, and multiple comparisons

- Provides a deeper understanding and appreciation of the history of statistics

- Discusses disagreements in the literature, which make for interesting reading

- Gives guidance on various aspects of statistics research by reading good examples in the literature

- Promotes the use of good English style in the presentation of statistical ideas by learning from well-written papers

- Includes an appendix of style tips on writing statistical papers

This book is aimed at researchers and graduate students in statistics and biostatistics, who are interested in the history of statistics and would like to deepen their understanding of seminal ideas and controversies. It could be used to teach a special topics course or useful for any researchers keen to understand the subject better and improve their statistical presentation skills.

Roderick J. A. Little is the Richard D. Remington Distinguished University Professor Emeritus at the University of Michigan, where he also holds emeritus appointments in the Department of Statistics and the Institute for Social Research. After secondary school at Glasgow Academy, he received a BA in mathematics from Gonville and Caius College, Cambridge University and MSc and PhD degrees in statistics from the Imperial College of Science and Technology, London University. Prof. Little is a pioneer and thought leader in the fields of statistical analysis with missing data and Bayesian inference in sample surveys and causal inference. He has received some of the highest honors in statistics and science, including being elected to the U.S. National Academy of Medicine and American Academy of Arts and Sciences.

Monographs on Statistics and Applied Probability

Editors: F. Bunea, R. Henderson, L. Levina, N. Meinshausen, R. Smith

For more information about this series please visit: https://www.crcpress.com/Chapman–HallCRC-Monographs-on-Statistics–Applied-Probability/book-series/CHMONSTAAPP

Seminal Ideas and Controversies in Statistics

Roderick J. A. Little

CRC Press
Taylor & Francis Group
Boca Raton London New York

CRC Press is an imprint of the
Taylor & Francis Group, an **informa** business
A CHAPMAN & HALL BOOK

Cover images (clockwise from top left): Donald B. Rubin, John W. Tukey, Ronald A. Fisher, Deborah G. Mayo, Bradley Efron, Jerzy Neyman.

First edition published 2025
by CRC Press
2385 NW Executive Center Drive, Suite 320, Boca Raton FL 33431

and by CRC Press
4 Park Square, Milton Park, Abingdon, Oxon, OX14 4RN

CRC Press is an imprint of Taylor & Francis Group, LLC

ISBN: 9781032497174 (hbk)
ISBN: 9781032493565 (pbk)
ISBN: 9781003395164 (ebk)

DOI: 10.1201/9781003395164

Typeset in Palatino
by KnowledgeWorks Global Ltd.

To Robin, David, and Andrew

Contents

Preface

Statistics has developed as a field through seminal papers and fascinating controversies. This book concerns a wide-ranging set of 15 statistical topics, grouped into three sets:

Part I, Chapters 1–6. Philosophical approaches to statistical inference,

Part II, Chapters 7–12. Statistical methodology, and

Part III, Chapters 13–15. Topics on statistical design, focusing on the role of randomization.

The chapters are grouped by topic and have some interconnections, but they are also freestanding and do not need to be read in order.

In each chapter, I list one or more key papers on these topics and include as other reading some later papers that may help to explain them. I then summarize the main ideas of these papers and give my personal perspective on them. The goal is to expand readers knowledge on statistics literature and encourage a historical perspective on the subject. I am not a historian, and my goal is understanding rather than historical accuracy. I also acknowledge my own limitations – I welcome hearing from readers who think the book has it wrong. Proofs are avoided – the focus is on ideas, not mathematical completeness – and simple examples are favored over generality.

The topics covered here have motivated my interest in statistics over my career. They include what I consider the most important statistics paper in the latter part of the 20th century (can you guess which one that is, after reading the book?). Also, we all love controversies – who has not turned to a contentious Royal Statistical Society discussion before reading the main paper? The list is far from comprehensive, and others would have undoubtedly chosen different topics, although I suspect there would be some overlap.

This book is intended for individuals familiar with the main tools of statistics like multiple regression, repeated-measures analysis, basic properties of distributions, and key asymptotic approaches such as maximum likelihood. The target audience includes doctoral students in statistics and biostatistics, and other statisticians who know the basics but are interested in the history of statistics and would like to deepen their understanding of key ideas and controversies.

When we teach a subject we generally say, "this is how to do it" without much discussion of weaknesses or alternatives approaches. This book is different in that my purpose is to present the clash of ideas, and hence to sow doubt and confusion about topics on the grounds that they may promote a deeper understanding. You will see I have a perspective on many of these topics, but you don't have to share the same views.

When taking coursework, students may not have read much of the original literature in statistics journals. Most of the papers discussed here are well written, if not always using modern-day English. Thus, I think reading the set of papers is helpful to gaining an appreciation for how to write well on statistics topics. This is particularly useful for students who will be writing doctoral dissertations, but clear writing and communication skills are important in many careers that involve statistics. In the appendix, I offer some style tips on how to write good statistical articles or other written communications involving statistics.

Key aims of the book are:

- To cover and discuss a number of important ideas in the history of statistics, concerning (a) philosophical approaches, (b) seminal problems in statistical analysis, and (c) design topics, focusing on the role of randomization
- Provide a deeper understanding and appreciation of the history of statistics
- Discuss disagreements in the literature, which make for interesting reading
- Learn various aspects of statistics research by reading good examples in the literature
- Promote good English style in the presentation of statistical ideas by learning from papers that are well written. My own style tips on writing statistical papers are included as an Appendix.

Topics are organized into the three areas as follows:

Part I: Statistical Inference

Chapter 1. Fisher and the Method of Maximum Likelihood

The paper:

Fisher, A. (1922a). On the mathematical foundations of theoretical statistics (with discussion). *Phil. Trans. Roy. Soc. London. Ser. A*, 222, 309–368.

Other reading:

Stigler, S. (2005). Fisher in 1921. *Statist. Sci.*, 20, 1, 32–49.

Seminal ideas: Method of maximum likelihood; sufficiency; consistency; efficiency; Fisher information.

The controversies: Maximum likelihood vs. method of moments; to Bayes or not to Bayes.

Chapter 2. To C or Not to C, That is the Question

The paper:

Yates, F. (1984). Tests of significance for 2 × 2 contingency tables (with discussion). *J. Roy. Statist. Soc., Ser. A*, 147, 426–463.

Other reading:

Hitchcock, D. B. (2009). Yates and contingency tables: 75 years later. *Electronic J. History Prob. Statist.*, 5, 2, https://www.jehps.net/decembre2009.html.

Howard, J. V. (1998). The 2 × 2 table: a discussion from a Bayesian viewpoint. *Statist. Sci.* 13, No. 4, 351–367.

Little, R. J. (1989). On testing the equality of two independent binomial proportions. *Amer. Statist.*, 43, 283–288.

Lyderson, S., Fagerland, M. W. & Laake, P. (2009). Tutorial in biostatistics: Recommended tests for association in contingency tables. *Statist. Med.* 28, 1159–1175.

Seminal idea: Ancillarity.

The controversies: Fisher's exact test vs. Pearson chi-squared test for independence in a (2 × 2) table; the role of ancillary statistics in frequentist inference.

Chapter 3. Frequentist Flaps: Hypothesis Testing, Significance Testing, or Something Else?

The papers:

Fisher, R. A. (1955). Statistical methods and scientific induction. *J. Roy. Statist. Soc., Ser. B*, 17, 1, 69–78.

Neyman, J. & Pearson, E. S. (1933). On the problem of the most efficient tests of statistical hypotheses. *Phil. Trans. Roy. Soc. London, Ser. A*, 231, 289–337.

Neyman, J. (1956). Note on an article by Sir Ronald Fisher. *J. Roy. Statist. Soc., Ser. B*, 18, 2, 288–294.

Wasserstein, R. L. & Lazar, N. A. (2016). The ASA's statement on p-values: context, process, and purpose. *Amer. Statist.*, 70, 2, 129–133, with supplemental comments at: https://www.tandfonline.com/doi/full/10.1080/00031305.2016.11541 08?cookieSet=1

Other reading:

Benjamin, D. J. et al. (2018). Redefine statistical significance. *Nat. Hum. Behav.*, 2, 1, 6–10.

Benjamini, Y., De Veaux, R., Efron, B., Evans, S., Glickman, M., Graubard, B. I., He, X., Meng, X.-L., Reid, N., Stigler, S. M., Vardeman, S. B., Wikle, C. K., Wright, T., Young, L. J., & Kafadar, K. (2021). ASA President's Task Force statement on statistical significance and replicability. *Harvard Data Science Review*, 3 (3).

Lehmann, E. L. (1993). "The Fisher, Neyman-Pearson theories of testing hypotheses: One theory or two?" *J. Amer. Statist. Assoc.* 88, 201–208.

McShane, B. B. & Gal, D. (2017). Statistical significance and the dichotomization of evidence. *J. Amer. Statist. Assoc.*, 112, 519, 885–895.

McShane, B. B, Gal, D., Gelman, A., Robert, C. & Tackett, J. L. (2019). Abandon statistical significance. *Amer. Statistician*, 73: suppl., 235–245.

Seminal ideas: Hypothesis testing; likelihood ratio tests.

The controversies: Neyman/Pearson vs. Fisher on hypothesis testing; role of hypothesis testing in statistical inference.

Chapter 4. Fiducial Inference and the Behrens-Fisher Problem

The papers:

Fisher, R. A. (1935b) The fiducial argument in statistical inference. *Ann. Eugenics*, 8, 391–398.

Ghosh, M. and Kim, Y.-H. (2001). The Behrens-Fisher problem revisited: a Bayes-frequentist synthesis. *Can. J. Statist.*, 29, 1, 5–17.

Welch, B. L. (1938) The significance of the difference between two means when the population variances are unequal. *Biometrika*, 29, 350–362.

Welch, B. L. (1956). Note on some criticisms made by Sir Ronald Fisher. *J. Roy. Statist. Soc. Ser. B*, 18, 2, 297–302.

Other reading:

Seidenfeld, T. (1992). R. A. Fisher's Fiducial argument and Bayes' Theorem. *Statist. Sci.*, 7, 3, 358–368.

Zabell, S. L. (1992). R.A. Fisher and the Fiducial argument. *Statist. Sci.* 7, 3. 369–387.

Seminal ideas: Pivotal quantities; fiducial inference.

The controversies: The soundness of Fisher's fiducial inference; alternative solutions of comparison of means from normal samples.

Chapter 5. Do You Like the Likelihood Principle?

The paper:

Birnbaum, A. (1962). On the foundations of statistical inference (with discussion). *J. Am. Statist. Assoc.*, 57, 269–326.

Other reading:
Berger, J. O. & Wolpert, R. L. (1988). *The Likelihood Principle*. Institute of Mathematical Statistics Lecture Notes-Monograph Series, 6, 1–199. Hayward, CA: Institute of Mathematical Statistics.

Mayo, D. G. (2014). On the Birnbaum argument for the strong likelihood principle (with discussion). *Statist. Sci.*, 29, 2, 227–266.

Seminal idea: The likelihood principle

The controversy: Validity of the likelihood principle and its implications.

Chapter 6. A Bayesian/Frequentist Compromise: Calibrated Bayes

The papers:

Box, G. E. P. (1980). Sampling and Bayes' inference in scientific modelling and robustness. *J. Roy. Statist. Soc. Series A*, 143, 4, 383–430.

Rubin, D. B. (1984). Bayesianly justifiable and relevant frequency calculations for the applied statistician. *Ann. Statist.*, 12, 4, 1151–1172.

Other reading:

Little, R. J. (2006). Calibrated Bayes: A Bayes/frequentist roadmap. *Amer. Statist.*, 60, 3, 213–223.

Seminal idea: Calibrated Bayes – a Bayes-frequentist compromise.

The controversy: Frequentist vs. Bayesian inference.

Part II. Statistical Methods

Chapter 7. Baseball Averages, Foreign Cars, and Shrinkage Estimation

The paper:

Efron, B. & Morris, C. (1977). Stein's paradox in statistics. *Sci. Amer.*, 1977, 119–127.

Other reading:

Efron, B. & Morris, C. (1973). Stein's estimation rule and its competitors—An empirical Bayes approach. *J. Amer. Statist. Assoc.*, 68, 341, 117–130.

Rubin, D. B. (1980). Using empirical Bayes techniques in the law school validity studies (with discussion). *J. Amer. Statist. Assoc.*, 75, 372, 801–881.

Shen, W. & Louis, T. A. (1998). Triple-goal estimates in two-stage hierarchical models. *J Roy. Statist. Soc. Ser B*, 60, 2, 455–471.

Seminal ideas: James-Stein Theorem; empirical Bayes methods.

The controversy: Statistics as mathematics or modeling.

Chapter 8. Alternatives to Least Squares in Regression

The paper:

Dempster, A. P., Schatzoff, M. & Wermuth, N. (1977). A simulation study of alternatives to ordinary least squares (with discussion). *J. Amer. Statist. Assoc.*, 72, 357, 77–106.

Other reading:

Chaibub Neto, E., Bare, J. C., Margolin, A. A. (2014). Simulation studies as designed experiments: the comparison of penalized regression models in the "large p, small n" setting. *PLoS ONE*, 9, 10, e107957.

Tibshirani, R. (1996). Regression shrinkage and selection via the lasso. *J. Roy. Statist. Soc. Ser. B*, 58, 1, 267–288.

Seminal ideas: Alternatives to least squares in situations where it performs poorly; role and design of simulation studies in statistics.

The controversy: Generality of conclusions of the Dempster et al. simulation study.

Chapter 9. Multiple Perspectives on Multiple Comparisons

The papers:

Benjamini, Y. & Hochberg, Y. (1995). Controlling the false discovery rate: a practical and powerful approach to multiple testing. *J. Roy. Statist. Soc. Ser. B*, 57, 1, 289–300.

Berry, D. (2012). Multiplicities in cancer research: ubiquities and necessary evils. *J. Nat. Cancer Inst.*, 104, 1124–1132.

Cox, D. R. (1965). A remark on multiple comparison methods. *Technometrics*, 7, 2, 223–224.

Gelman, A., Hill, J. & Yajima, M. (2012). Why we (usually) don't have to worry about multiple comparisons. *J. Res. Educ. Effectiveness*, 5, 189–211.

Rothman, K. J. (1990). No adjustments are needed for multiple comparisons. *Epidemiology*, 1, 43–46.

Tukey, J. W. (1991). The philosophy of multiple comparisons. *Statist. Sci.*, 6, 1, 100–116.

Seminal ideas: Methods for controlling type 1 errors in testing; confidence coverage in confidence intervals; false discovery rate.

The controversy: Whether and how to handle multiple comparisons.

Chapter 10. Generalized Estimating Equations

The paper:

Liang, K.-Y. & Zeger, S. L. (1986). Longitudinal data analysis using generalized linear models. *Biometrika*, 73, 1, 13–22.

Other reading:

Hubbard, A. E., Ahern, J., Fleischer, N. L., Van Der Laan, M., Lippman, S. A., Jewell, N., Bruckner, T., & Satariano, W. A. (2010). To GEE or not to GEE. Comparing population average and mixed models for estimating the associations between neighborhood risk factors and health. *Epidemiology*, 21, 467–474.

Pepe, M. S. & Anderson, G. L. (1994). A cautionary note on inference for marginal regression models with longitudinal data and general correlated response data. *Commun. Statist. – Simul. Comp.*, 23, 4, 939–951.

Seminal idea: Generalized estimating equations for longitudinal data analysis.

The controversy: Generalized estimating equations vs. likelihood-based methods.

Chapter 11. The Bootstrap and Bayesian Monte-Carlo Methods

The papers:

Efron, B. (1979). Bootstrap methods: another look at the jackknife. *Ann. Statist.*, 7, 1, 1–26.
Gelfand, A. E. & Smith, A. F. M. (1990). Sampling-based approaches to calculating marginal densities, *J. Amer. Statist. Assoc.*, 85, 410, 398–409.

Other reading:

DiCiccio, T. J. & Efron, B. (1996). Bootstrap confidence intervals. *Statist. Sci.*, 11, 3, 189–228.
Rubin, D. B. (1981). The Bayesian Bootstrap. *Ann. Statist.* 9, 1, 130–134.
Tanner, M. A. & Wong, W. H. (1987). The calculation of posterior distributions by data augmentation. *J. Am. Statist. Assoc.*, 82, 398, 528–540.

Seminal ideas: The bootstrap; Bayesian Markov Chain Monte-Carlo computation.

Chapter 12. Exploratory Data Analysis and Data Science

The papers:

Tukey, J. W. (1962). The future of data analysis. *Ann. Math. Statist.*, 33, 1, 1–67.
Breiman, L. (2001). Statistical modeling: two cultures. *Statist. Sci.* 16, 3, 199–231.

Other reading:

Donoho, D. (2017). 50 years of data science. *J. Comp. Graphical Statist.*, 26, 4, 745–766.
Mitra, N., (2021). Introduction to Special Issue: Commentaries on Breiman's Two Cultures paper. *Observational Studies*, 7, 1, 1–2, and the other papers in that volume.
van der Laan, M. J., Polley, E. C., & Hubbard, A. E. (2007). "Super learner". *Statist. Applic. Genetics Mol. Biol.*, 6, 1, Article 25.

Seminal ideas: Exploratory data analysis and data science; assessment of algorithms via prediction.

The controversies: Algorithmic methods vs. classical statistical models; the meaning of data science.

PART III: Topics in Design

Chapter 13. Randomization in Survey Sampling

The paper:

Neyman, J. (1934). On the two different aspects of the representative method: the method of stratified sampling and the method of purposive selection. *J. Roy. Statist. Soc.*, 97, 4, 558–625.

Other reading:

Little, R. J. (2012). Calibrated Bayes: An alternative inferential paradigm for official statistics (with discussion and rejoinder). *J. Official Statist.*, 28, 3, 309–372.

Little, R. J. (2014). Survey sampling: past controversies, current orthodoxies, and future paradigms. In *Past, Present and Future of Statistical Science*, X. Lin, D.L. Banks, C. Genest, G. Molenberghs, D. W. Scott, & J.-L. Wang, eds. CRC Press.

Rubin, D. B. (1976). Inference and missing data. *Biometrika*, 63, 581–592.

Seminal ideas: Definition of confidence interval; stratified random sampling; Neyman allocation.

The controversies: Importance of random sampling; design-based vs. model-based survey inference.

Chapter 14. Randomized Clinical Trials and the Neyman/Rubin Causal Model

The papers:

Medical Research Council (1948). Streptomycin treatment of pulmonary tuberculosis: a Medical Research Council investigation. *Brit. Med. J.*, 2, 769–782.

Rubin, D. B. (1978). Bayesian inference for causal effects: The role of randomization. *Ann. Statist.*, 6, 1, 34–58.

Other reading:

Little, R. J. & Lewis, R. J. (2021). Estimands, estimators and estimates. *J. Amer. Med. Assoc.*, 326, 10, 967–968.

Seminal ideas: Neyman/Rubin causal model; the role of randomization in studies comparing treatments.

The controversy: The importance of randomization.

Chapter 15. Propensity Score Methods

The paper:

Rosenbaum, P. R. & Rubin, D. B. (1983). The central role of the propensity score in observational studies for causal effects. *Biometrika*, 70, 1, 41–55.

Other reading:

Little, R. J. (2022). Some reflections on Rosenbaum and Rubin's propensity score paper. *Observational Studies*, 9, 1, 69–75.

Seminal ideas: Propensity score methods for observational studies comparing treatments; selection bias and nonresponse.

Appendix: Twenty Style and Grammar Tips for Statistics Writing

Acknowledgments

It is a pleasure to thank many colleagues who have helped me with this project. The book is based on a doctoral-level course I have taught six times at the University of Michigan, and the enthusiasm and myriad contributions of the students and faculty kibitzers over the years motivated me to write the book and helped to shape the material. In particular, Bhramar Mukherjee enthusiastically supported the concept of the course, co-taught it with me on one occasion, and contributed her deep insights and love of the history of statistics. Two doctoral students, Andrew Beck from biostatistics and Gabriel Durham from statistics, kindly volunteered to read and comment on a first draft of the book and provided many valuable insights and edits, leading to significant improvements in the text (the remaining errors are all mine) Philip Boonstra, Andrew Gelman, Jack Kalbfleisch, Andrew Little, Deborah Mayo, Stephen Stigler, Trivellore Raghunathan, Donald Rubin, and two anonymous reviewers reviewed all or parts of the book or made constructive suggestions about content. I apologize to others I may have missed. Finally, Rob Calver at Chapman & Hall/CRC has been a highly supportive editor, and many thanks also to Sherry Thomas for editorial assistance.

Rod Little
Ann Arbor, Michigan
August 3, 2024

Part I

Statistical Inference

1

Fisher and the Method of Maximum Likelihood

The paper:

Fisher, R. A. (1922a). On the mathematical foundations of theoretical statistics (with discussion). *Phil. Trans. Roy. Soc. London. Ser. A*, 222, 309–368.

Other reading:

Stigler, S. (2005). Fisher in 1921. *Statist. Sci.*, 20, 1, 32–49.

1.1 Introduction

Fisher (1922a) was certainly not the first important paper in the history of statistics – major earlier contributors include Laplace, Quetelet, Gauss, Bayes, Galton, and Karl Pearson, to name a few. However, Fisher is often considered the founder of modern statistical methods, and there is no doubt that the ideas in this paper exerted an enormous influence on the field. It is remarkable how modern this paper feels, written as it was some 100 years ago.

After dyspeptic comments about the state of statistical theory and insightful comments about what the field of statistics is about, Fisher introduces the likelihood function and the maximum likelihood (ML) estimate and establishes its key properties. Specifically, a parametric statistical model for data y involves choosing a probability density $f_Y(y \mid \theta)$ indexed by unknown parameters θ. The likelihood function is simply the probability density treated as a function of the parameters θ rather than the data y, that is:

$$L(\theta \mid y) \propto f(y \mid \theta), \tag{1.1}$$

where the proportionality constant can depend on y but not on θ. The ML estimate, say $\hat{\theta}$, is the value of θ that maximizes the likelihood; for many important statistical models, it is unique.

Fisher also defines other key concepts in mathematical statistics, including consistency and efficiency of an estimate, sufficiency, and Fisher information. It's an impressive list for a single paper, particularly given that the method of ML remains one of the basic tools of modern statistical inference.

Stigler's (2005) fascinating article on Fisher's early years is suggested as other reading. He sums up Fisher (1922a) as follows:

> The paper is an astonishing work: it announces and sketches out a new science of statistics, with new definitions, a new conceptual framework and enough hard mathematical analysis to confirm the potential and richness of this new structure.

Stigler describes Fisher's early influences and how his correspondence with Karl Pearson, then the editor of *Biometrika*, prompted the development of Fisher's (1922) paper. Stigler notes that Fisher's fully formed statistical theory came "out of the blue," as prior to the time of writing, he had focused on commentaries on eugenics and problem-solving particular statistics questions.

1.2 Main Points of Fisher's Paper

In the opening section, Fisher comments caustically on the confused state of statistical theory at that time, laying blame for the confusion on the rise of Bayesianism. To quote from the paper:

> …in statistics a purely verbal confusion has hindered the distinct formulation of statistical problems; for it is customary to apply the same name, mean, standard deviation, correlation coefficient, etc., both to the true value which we should like to know, but can only estimate, and to the particular value at which we happen to arrive by our methods of estimation; so also in applying the term probable error, writers sometimes would appear to suggest that the former quantity, and not merely the latter, is subject to error. It is this last confusion, in the writer's opinion, more than any other, which has led to the survival to the present day of the fundamental paradox of inverse probability, which like an impenetrable jungle arrests progress towards precision of statistical concepts… though we may agree wholly with CHRYSTAL that inverse probability is a mistake … there yet remains the feeling that such a mistake would not have captivated the minds of LAPLACE and POISSON if there had been nothing in it but error.

Fisher makes an important distinction here between the population parameter – the estimand – and the parameter estimate. He then builds on it by criticizing the "mistaken" Bayesian practice of assigning a distribution to the estimand rather than to the estimate – from his frequentist perspective, the estimand is a fixed number, so how can it have a distribution? He grudgingly concedes, however, that a theory embraced by the giants like Laplace and Poisson might be worth something.

In Section 2, Fisher describes the main purpose of statistical methods as the reduction of data, motivating his later definition of sufficient statistics. He then formulates the "law of distribution" for data, assumed to be drawn from a hypothetical population; we might term this the "statistical model" in more modern language. His account of the probability distributions of statistics from repeated sampling follows standard frequentist lines. He concludes with a passing reference to the appropriate reference set for frequentist calculations, or in his words "what is, and what is not, relevant in repeated sampling." He mentions as an example the case of whether to fix margins in a 2×2 contingency table, which is the main topic of the next chapter.

In Section 3, Fisher distinguishes three problems in the reduction of data:

1. Problems of specification, or what I would label the choice of statistical model.

2. Problems of estimation, that is, choice of statistics to estimate parameters in the statistical model.

3. Problems of distribution, that is, the sampling distribution of estimates of parameters in the assumed model.

These distinctions remain very relevant today. I note the absence of what we might term "statistical inference," that is the assessment of uncertainty through hypothesis testing or interval estimation. In particular, Bayesian posterior distributions are beyond the pale, and fiducial and confidence intervals came later (see Chapters 3 and 4). Large-sample standard errors of estimates – what Fisher called "probable errors" – are discussed, but these are used to determine the efficiency of point estimates, rather than for interval estimation, as in confidence or posterior credible intervals. Also, the validity of the statistical model is not in question here, with no consideration of model checks or what we would now call the robustness of estimates to departures from model assumptions.

Section 4 is remarkable for defining three key concepts that are still central to modern statistics, namely consistency, efficiency, and sufficient statistics. To quote from the paper:

> The common-sense criterion employed in problems of estimation may be stated thus: That when applied to the whole population the derived statistic should be equal to the parameter. This may be called the **Criterion of Consistency**...
> ... Consideration of the above example will suggest a second criterion, namely:- That in large samples, when the distributions of the statistics tend to normality, that statistic is to be chosen which has the least probable error. This may be called the **Criterion of Efficiency**.
> ... The criterion of efficiency is still to some extent incomplete ...the complete criterion suggested by our work on the mean square error is: That the statistic chosen should summarise the whole of the relevant information supplied by the sample. This may be called the **Criterion of Sufficiency**.

Fisher illustrates these ideas for the problem of estimating the variance of a normal distribution. He shows that the mean squared error is an efficient estimate, and the mean absolute error is not. The efficiency of the mean squared error is related to the fact that it is a sufficient statistic for the variance, and the relationship between efficiency and sufficiency is elucidated in an elegant analysis of the asymptotic normal distribution of two estimates, one of which is sufficient. Stigler (2005) discusses how Fisher's analysis of the normal model motivated his more general definition of a sufficient statistic.

Section 5 nominally concerns examples of consistency, focusing on the method of moment estimation. The most popular class of models at that time was the system of probability distributions devised by Karl Pearson. For once, Fisher is complimentary about Pearson's work:

> We may instance the development by PEARSON of a very extensive system of skew curves, the elaboration of a method of calculating their parameters, and the preparation of the necessary tables, a body of work which has enormously extended the power of modern statistical practice, and which has been, by pertinacity and inspiration alike, practically the work of a single man... of even greater importance is the introduction of an objective criterion of goodness of fit.

Pearson's approach to estimating parameters was the method of moments, where parameter estimates for a distribution with k parameters are obtained by equating the first k population and sample moments. Fisher concedes that the method was "without question of practical utility," but then notes that it does not necessarily yield efficient estimates and that there are situations where the method fails. This sets the stage for the next section, where Fisher proposes ML as an alternative estimation approach.

As an example where the method of moments fails, Fisher considers estimating the location parameter m from a random sample from the Cauchy distribution with density

$$f(x \mid m) = \frac{1}{\pi\left(1 + (x - m)^2\right)}.$$

The method of moments would equate the sample mean to the population mean, and that does not work here because this distribution does not have a mean.

The heart of the paper lies in Section 6, entitled "Formal Solution of Problems of Estimation." He defines the likelihood and distinguishes it from probability, using the estimation of a binomial proportion as illustration:

> ...I must indeed plead guilty in my original statement of the Method of the Maximum Likelihood (9) to having based my argument upon the principle of inverse probability... Upon consideration, therefore, I perceive that the word probability is wrongly used in such a connection ...

> We must return to the actual fact that one value of p, of the frequency of which we know nothing, would yield the observed result three times as frequently as would another value of p. If we need a word to character-ise this relative property of different values of p, I suggest that we may speak without confusion of the likelihood of one value of p being thrice the likelihood of another, bearing always in mind that likelihood is not here used loosely as a synonym of probability, but simply to express the relative frequencies with which such values of the hypothetical quantity p would in fact yield the observed sample.

Fisher then proposes the ML estimate as a general solution of the estima-tion problem, illustrating the point by deriving the ML estimate of the prob-ability for a binomial sample as the sample proportion. He also derives the inverse of the Fisher information as the large sample variance of the ML esti-mate, for a univariate problem, sketching the extension to multiparameter problems in Section 7.

Fisher shows that the ML estimate does not lose information about param-eters and must be a function of the minimal sufficient statistic. However, is the ML estimate by itself necessarily sufficient? Fisher seemed to think so, but he hedged on this question; in Section 6, he writes:

> For the solution of problems of estimation we require a method which for each particular problem will lead to automatically to the statistic by which the criterion of sufficiency is satisfied. Such a method is, I believe, provided by the Method of Maximum Likelihood, although I am not sat-isfied as to the mathematical rigour of any proof which I can put forward to that effect.

Fisher's supposition is wrong – the ML estimate is not necessarily suffi-cient. For example, the ML estimate of m in the Cauchy location example described above is not a sufficient statistic for m; the minimal sufficient sta-tistic is the set of order statistics.

A large part of Section 6 concerns the invariance property of ML under transformations – that the ML estimate $\hat{\phi}$ of a function of a parameter θ, say $\phi = f(\theta)$, is the function evaluated at the ML estimate $\hat{\theta}$ of θ, that is $\hat{\phi} = f(\hat{\theta})$. Fisher then attacks the Bayesian approach because uniform priors on param-eters do not have a similar invariance property. He writes:

> [Bayes' Theorem] would, if true, be of great importance in bringing an immense variety of questions within the domain of probability. It is, how-ever, evidently extremely arbitrary. Apart from evolving a vitally important piece of knowledge, that of the exact form of the distribution of values of p, out of an assumption of complete ignorance, it is not even a unique solution.

The remainder of Fisher's paper, more than half of its total length, con-sists mainly of a detailed analysis of the relative efficiency of the method of moments for a variety of Pearson-type distributions, including the Pearson Type III or gamma distribution (Section 8), location and scale distributions (Section 9), general Pearson-type curves (Section 10), and discontinuous distributions, including the Poisson and grouped normal (Section 12). The conceptual breakthroughs occur in the first half of the paper, but in the second half, Fisher demonstrates his great mathematical acumen. To quote Stigler (2005):

> The labor in producing this investigation must have been immense, but it had a satisfying payoff: the method of moments was shown to only have high efficiency when the curve was near the normal curve (where it was fully efficient, in fact maximum likelihood). For other cases Pearson's method could perform abysmally. For Type III curves (Gamma and chi-squared densities), Fisher quoted efficiencies for low degrees of freedom dropping off from 0.2727 to 0.

This detailed analysis of the deficiencies of the method of moments would not have pleased Pearson, and Stigler (2005) describes the simmering animosity between Fisher and Pearson that Fisher's paper aroused.

1.3 Discussion

Clearly, Fisher (1922a) is a blockbuster – the ideas in this paper, together with Fisher's work on experimental design and analysis of variance, can be seen as the two monumental pillars supporting Fisher's status as the founder of modern statistics. But naturally, the paper does not solve everything – the focus on estimation rather than inference is limiting, consequences of model misspecification are not addressed, and the optimal properties of ML are asymptotic and not unique to ML. Thus, I would say Fisher hit on a superb solution, but perhaps other solutions also demand attention.

1.3.1 Maximum Likelihood versus Method of Moments

Fisher's paper can be seen as the first salvo in a long clash between two major philosophies of statistical inference:

1. Pearson's method of moments, which led to generalized estimating equations (GEEs), as discussed in the repeated-measures context in Chapter 10; and
2. Fisher's ML, variants such as Bayesian inference, and extended likelihood approaches such as conditional, marginal, or partial likelihood.

Fisher's paper implies a clear victory for ML, but the winner is not quite so clear-cut when issues of model misspecification are taken into account. ML is fully efficient but requires a fully and correctly specified statistical model; the method of moments (and extensions such as generalized estimating equations, GEE) are less efficient but may be less reliant on a specific parametric model, and hence are potentially more robust.

Fisher illustrated the superiority of ML over the method of moments for the example of a sample from the Cauchy distribution, which is the t distribution with $v = 1$ degree of freedom. Consider now inference about the mean μ and scale parameter σ of the t distribution with known degrees of freedom v, where $v > 2$. (Note in what follows that the variance of this distribution is not σ^2 but $\sigma^2/(1-2/v)$.) The density for a random sample $(x_i, i = 1,...,n)$ is:

$$\prod_{i=1}^{n} f(x_i \mid \mu,\sigma^2), \text{ where } f(x_i \mid \mu,\sigma^2) = c/\left(1+(x_i-\mu)^2/(v\sigma^2)\right)^{(v+1)/2},$$

and c is a known function of v, omitted to avoid clutter. The likelihood is the expression regarded as a function of (μ,σ^2). The ML estimates can be obtained by iteratively reweighted least squares: given current parameter estimates $(\mu^{(t)}, \sigma^{(t)})$ of (μ,σ^2), the $(t+1)$th iteration computes new estimates as:

$$\mu^{(t+1)} = \sum_{i=1}^{n} w_i^{(t)} x_i / \sum_{i=1}^{n} w_i^{(t)}; \hat{\sigma}^{(t+1)2} = \frac{1}{n}\sum_{i=1}^{n} w_i^{(t)}\left(x_i - \hat{\mu}^{(t+1)}\right)^2,$$

where the weight is

$$w_i^{(t)} = E(w_i \mid x_i, \theta^{(t)}) = \frac{v+1}{v+d_i^{(t)2}}, d_i^{(t)} = (x_i - \mu^{(t)})/\sigma^{(t)}.$$

Thus, observations are downweighted by an amount $w_i^{(t)}$ that depends on v and the standardized distance of each x_i from the current mean estimate $\mu^{(t)}$. The approach is an example of an EM algorithm (Dempster, Laird & Rubin 1977) for a random sample from a normal model with variance σ^2/u_i, where u_i is a latent (unobserved) variable distributed as chi-squared with v degrees of freedom. See, for example, Example 8.4 in Little and Rubin (2019) for more details.

The method of moments estimates of μ and σ^2 are simply the unweighted sample mean and appropriately scaled multiple of the sample variance:

$$\tilde{\mu} = \sum_{i=1}^{n} x_i/n, \tilde{\sigma}^2 = (1-2/v)\sum_{i=1}^{n}(x_i - \tilde{\mu})^2/n.$$

Which of these sets of estimates is better? The ML estimates are more efficient but rely more heavily on the assumption of the t model with known v. The method of moments is less efficient but yields consistent estimates under weaker assumptions. Interestingly, downweighting outliers (as in the ML approach) is often thought of as a form of robust inference, but in an important sense, it is *less* robust than the method of moments; the weight of observations in each tail that are the same distance from the estimated mean is the same, implying an assumption of symmetry of the underlying distribution; for asymmetric distributions, the weighting yields an inconsistent estimate of the mean, but the (unweighted) method of moments estimate remains consistent.

1.3.2 Optimality of Maximum Likelihood, and the Bayes Question

The relationship between ML and sufficient statistics, and the consistency and efficiency of ML under a correctly specified model, are important findings. ML gives a unique answer providing the parameters are identified, but it is not the *only* procedure that is efficient in large samples. In fact, Bayesian inference, which Fisher derides, also delivers efficient estimates, provided the prior distribution's support includes the true value of the parameter. Bayesian inference plays a prominent role in many of the chapters in this book, particularly Chapter 6. Its basic features are as follows.

Bayesian inference for the model yielding the likelihood (1.1) adds a prior distribution $\pi(\theta)$ for the parameters θ and bases inferences on the resulting posterior distribution:

$$p(\theta \mid y) = c\pi(\theta) \times L(\theta \mid y), \tag{1.2}$$

which is a consequence of Bayes rule. Here, the constant c is chosen so that $\int p(\theta \mid y)d\theta = 1$. Examples of Bayes estimates are the mean, median, or mode of the posterior distribution. A $100(1-\alpha)\%$ posterior credible interval is an interval that includes $100(1-\alpha)\%$ of the probability in the posterior distribution.

Bayes and ML are closely related. Both bring in the data through the likelihood function. For a uniform prior $\pi(\theta) = $ const., the ML estimate is also the mode of the posterior distribution. Also, ML can be viewed as a form of large sample Bayes – asymptotically, the prior distribution $\pi(\theta)$ is dominated by the likelihood function in (1.2) and determines the posterior distribution.

ML is excellent if the model is correct and the sample size is large, but it has deficiencies in small samples, which Bayesian inference can correct with a judicious choice of prior. In particular, for normal models, Bayes with a Jeffreys' prior distribution yields a multivariate t posterior distribution, providing posterior credible intervals that incorporate degrees of freedom and t corrections, and posterior credible intervals for location parameters that are superior to ML in small samples. ML for a binomial proportion yields the Wald confidence interval, which behaves poorly in small samples, particularly for proportions lying close to zero or one. In this setting, Bayes with a weak prior distribution on the proportion gives much better frequentist answers (see, e.g., Brown, Cai & DasGupta 2001).

ML for the normal multiple linear regression with large numbers of associated predictors can be greatly improved by an informative prior on the parameters, as for example with ridge regression or the lasso – see Chapter 8 for more details.

Regarding Fisher's initial objection to Bayes in Section 1, I have never been convinced by the frequentist idea that "fixed parameters" cannot be assigned a distribution – for example, I have never understood the frequentist distinctions between "fixed" and "random" parameters in the analysis of variance models, as discussed more in Chapter 7. In Bayesian analysis, the key distinction is not between what is *fixed* and what is *random*, but rather what is *known* and what is *unknown*; prior and posterior distributions are used to quantify uncertainty in what is unknown, before and after seeing the data.

Fisher sought a single "optimal" answer to estimation, but I see the fact that Bayes produces a wealth of answers depending on the choice of prior distribution as a strength rather than a weakness. A problem with the best frequentist alternative to Bayesian credibility intervals – Neyman's confidence intervals – is that no answers exist for many statistical models, in the sense that there are no intervals with exact nominal confidence coverage for all values of the unknown parameters. An example of such a model is given in the discussion of the Behrens-Fisher problem in Chapter 4. Fisher rejects Bayes, but later thought he had solved the "problem of inverse probability" with the method known as fiducial inference, also discussed in Chapter 4.

Fisher correctly notes that uniform priors do not remain uniform after a nonlinear transformation of the parameters. Thus, Bayes combined with a uniform prior leads to inferential inconsistencies. However, draws from a posterior distribution – as in Markov-Chain Monte Carlo methods – do share with ML a property of transformation invariance, which can be of great utility in applications. That is, a draw from the posterior distribution of a nonlinear function ϕ of a parameter θ, say $\phi = f(\theta)$, is the function evaluated at a draw $\theta^{(d)}$ from the posterior distribution of θ, that is $\phi^{(d)} = f(\theta^{(d)})$. I'll say more about that in Chapter 11.

Suppose, for example, we are interested in inference about the principal components of the covariance matrix Σ of a set of variables, from a random sample of modest size, and we assume a multivariate normal model for the data. The ML estimates are the principal components of the sample covariance matrix S, and interval estimates can be obtained assuming large samples. A standard Bayesian solution assumes a conjugate inverse-Wishart prior distribution for Σ. The posterior distribution of Σ is then also inverse-Wishart, and it is a simple matter to obtain a draw $\Sigma^{(d)}$ from this distribution. The principal components of $\Sigma^{(d)}$ then provide a draw from their posterior distribution, and repeated draws can be used to simulate the posterior credibility intervals for the principal components. The Bayesian solution looks like ML in large samples but propagates uncertainty better in small or moderate samples.

1.4 Some Thought Questions on This Chapter

1. What in your opinion are the three most important reasons why this is a landmark paper in the history of statistics?

2. What are the three main problems of statistics, according to Fisher? Do you think this analysis is still valid today, 95 years later?

3. What was the primary approach to statistical estimation before Fisher's paper, as advocated by Karl Pearson and others? Outline the example Fisher uses where this approach fails but ML succeeds.

4. What properties of ML does Fisher advance in support of the method?

5. What example does Fisher use to suggest that ML is different from, and superior, to the Bayesian approach? Do you agree with his conclusion?

From Fisher's sweeping ML theory of inference for large samples, I now turn in Chapter 2 to a small sample inference problem involving just four numbers. The dataset is meager, but the ideas are profound, as we shall see.

2

To C or Not to C, That is the Question

The paper:

Yates, F. (1984). Tests of significance for 2×2 contingency tables (with discussion). *J. Roy. Statist. Soc., Ser. A*, 147, 426–463.

Other readings:

Hitchcock, D. B. (2009). Yates and contingency tables: 75 years later. *Electronic J. History Prob. Statist.*, 5, 2, 1–14.

Howard, J. V. (1998). The 2×2 table: a discussion from a Bayesian viewpoint. *Statist. Sci.* 13, 4, 351–367.

Little, R. J. (1989). On testing the equality of two independent binomial proportions. *Amer. Statist.*, 43, 283–288.

Lyderson, S., Fagerland, M. W. and Laake, P. (2009). Tutorial in biostatistics: recommended tests for association in contingency tables. *Statist. Med.*, 28, 1159–1175.

2.1 Introduction

Yates' (1984) paper is fascinating on many levels. It concerns a test that is universally taught in first statistics courses, the chi-squared test of independence in 2×2 contingency tables. When I applied for a visa to the United States, the immigration officer noted that I claimed to be a statistician and checked my veracity by grilling me about this chi-squared test!

The question concerns the appropriate choice of hypothesis test for the problem and raises fundamental issues of statistical inference. There are famous statisticians on both sides of the argument, so it's not surprising that the answer is not clearcut. Yates (1984) and the ensuing discussion generally favors the Fisher exact test and Yates' approximation of that method, but opinions are still divided – see, for example, the more recent review by Lyderson, Fagerland and Laake (2009). For me, the paper exposes some cracks in the armor of the frequentist approach to statistical inference, specifically, whether to condition on ancillary and approximately ancillary statistics. This issue recurs in a number of papers we discuss later, including Birnbaum's paper on the likelihood principle (Chapter 5) and Box and Rubin's papers on calibrated Bayes (Chapter 6). So, it sets the stage for later discussions, in the context of a basic statistics problem.

DOI: 10.1201/9781003395164-3

In my first course on statistics, taught by no less an eminence than D. R. Cox, I learnt to test the null hypothesis of independence using the Pearson chi-squared test for contingency tables with Yates' (1934) continuity correction. Yates was a follower of R.A. Fisher, and his successor as the head statistician at Rothamsted Experimental Station. He was an influential figure in statistics in the mid-20th century. The paper featured here was published some 50 years after his original 1934 paper, when he was over 80 years old, a tribute to his longevity.

Concerning the other readings, Hitchcock's paper contains some biographical details on Yates and celebrates the 75th anniversary of his 1934 paper.

Lyderson, Fagerland and Laake (2009) provide a clear review of the alternative approaches, reaching a different conclusion from that in Yates' (1984) paper. Howard (1998) provides a Bayesian perspective on the problem, discussed in Section 2.2.5 below. Little (1989) focuses on the philosophical issues raised by the controversy covered in Yates' 1984 paper and succeeding discussion, and this is also an emphasis of the summary here.

On a personal note, I met Yates, a charming man, when I was a young graduate student who knew next to nothing about statistics; we discussed the joys of traversing the Cuillin Ridge in Skye.

2.2 Description of the Problem

2.2.1 The Data

The data consist of the counts from a 2×2 contingency table with binary row variable A and binary column variable B, displayed in Table 2.1.

An important feature of the problem is whether margins of the table are fixed in repeated sampling under the design, or vary, that is, are random variables. The data in Table 2.1 could arise from designs that fix none, one, or both these margins:

TABLE 2.1

Notation for (2×2) Contingency Table

		B		Total	Proportions
		1	2		
A	1	x	$n_1 - x$	n_1	$p_1 = x/n_1$
	2	y	$n_2 - y$	n_2	$p_2 = y/n_2$
Total		$m_1 = x + y$	$m_2 = N - x - y$	$N = n_1 + n_2$	$p = m_1/N$

a. **No margins fixed design:** Suppose a random sample of size N is selected (or is assumed to have been randomly selected) from the population, and Table 2.1 is a cross-classification of the sample by two sampled variables A and B. The counts in both the margins are then random, and a natural model for the counts in the table is a multinomial distribution with index N and joint probabilities $\pi_{jk} = \Pr(A = j, B = k)$. I discuss this design in Section 2.2.4, for inference about the conditional probability that $B = 1$ in the subgroups with $A = 1$ and $A = 2$.

b. **One margin fixed design:** A common design in clinical trials fixes one margin, as follows. Let A denote a binary variable for two treatments $A = 1$ and 2, and suppose an independent sample of fixed size n_j is assigned treatment $A = j$, for $j = 1, 2$. Values of a binary outcome variable B, say $B = 1$ or 2, are then recorded for units in each sample. For treatment $A = 1$, say x units have $B = 1$, $n_1 - x$ units have $B = 2$, and $p_1 = x/n_1$. For treatment $A = 2$, y units have $B = 1$, $n_2 - y$ units have $B = 2$, and $p_2 = y/n_2$ (see Table 2.1).

   For this design, the A- margin (n_1, n_2) is fixed, and the B - margin (m_1, m_2) is random. The counts x and y are $\mathrm{Bin}(n_1, \pi_1)$ and $\mathrm{Bin}(n_2, \pi_2)$, respectively, where $\mathrm{Bin}(n, \pi)$ denotes the binomial distribution with count n and probability π. The joint distribution of x and y is:

$$\Pr(x, y \mid \pi_1, \pi_2, n_1, n_2) = \binom{n_1}{x} \pi_1^x (1 - \pi_1)^{n_1 - x} \binom{n_2}{y} \pi_2^y (1 - \pi_2)^{n_2 - y}, \quad (2.1)$$

where $\binom{n}{k} = \frac{n!}{k!(n-k)!}$. This distribution is also obtained from the multinomial distribution for the joint counts in design a, by conditioning on n_1 and n_2.

c. **Both margins fixed design:** The design of the famous tea-tasting experiment in Fisher (1935) leads to a table with both the row and column margins fixed. Given a cup of tea with milk to drink, a woman claimed to be able to tell whether the tea was added first ($A = 1$) or the milk was added first ($A = 2$). Fisher gave her eight cups of tea in a random order and told the woman that there were four of each variety. He asked her to identify for each cup if the tea or the milk was added first. The outcome variable B denotes whether she was right ($B = 1$) or wrong ($B = 2$). Evidently for this design, both margins are fixed and take the values $n_1 = n_2 = 4$ and $m_1 = m_2 = 4$.

The situation where both margins are fixed also arises for Fisher's randomization test, where the outcomes under the two treatments are fixed and the indicator for which treatment is assigned is the random variable. In this setting, both margins are fixed under the sharp null hypothesis that the outcomes for each individual in the study are the same under both treatments.

2.2.2 The Problem

The problem of interest is to assess whether A and B are associated. Let π_j denote the probability that $B = 1$ in the subpopulation with $A = j$ ($j = 1, 2$). The classical Neyman-Pearson hypothesis testing approach tests the null hypothesis of independence between the rows and column variables, or equivalently equality of π_1 and π_2:

$$H_0: \pi_1 = \pi_2,$$

against the one-sided alternative hypothesis

$$H_1: \pi_1 > \pi_2,$$

or the two-sided alternative hypothesis

$$H_2: \pi_1 \neq \pi_2.$$

The P-value for the test consists of the probability of getting a value of a test statistic (say T) as or more extreme than that observed, if the null hypothesis is true. Thus, for the one-sided test, the P-value is:

$$P = \Pr(T \geq t_{obs} \mid H_0),$$

where t_{obs} is the observed value of the test statistic. In classical Neyman-Pearson theory, the null hypothesis "rejected" in a test of size α if $P < \alpha$ and "accepted" (or better, "not rejected") if $P > \alpha$.

The key statistical issue here is the set of possible tables in repeated sampling that should be included in computing the P-value – what Fisher called the "reference set" for frequentist calculations. This reference set, and hence the P-value, differs depending on whether zero, one, or two margins are fixed in repeated sampling. A common belief is that the reference set is determined by the design – the margins that are fixed in the chosen design should be the margins that are fixed in the reference set. This philosophy leads to different reference sets for the three designs in Section 2.2.1. It also leads to conceptual problems, as described in the examples in Section 2.2.4.

2.2.3 The Pearson, Barnard, Fisher, and Yates Tests

I discuss here four of the many possible tests of independence for the (2×2) table that have been proposed.

a. **Pearson chi-squared test (P):** A point estimate of $\pi_1 - \pi_2$ is the difference in sample proportions $p_1 - p_2$. Dividing this estimate by its estimated standard error under H_0, namely $\sqrt{p(1-p)(1/n_1 + 1/n_2)}$ where p is the pooled proportion m_1/N, yields the following test statistic for testing H_0, the standardized difference in proportions:

$$Z_{Pe} = \frac{p_1 - p_2}{\sqrt{p(1-p)(1/n_1 + 1/n_2)}} = \frac{x(n_2 - y) - (n_1 - x)y}{\sqrt{m_1 m_2 n_1 n_2 / N}}. \tag{2.2}$$

   Assuming large samples, Z_{Pe} has a standard normal distribution under H_0 and the one-sided test for this test statistic has the P-value $1 - \Phi(z_{Pe,obs})$, where $z_{Pe,obs}$ is the observed value of the test statistic Z_{Pe}, and $\Phi()$ is the cumulative density function of the standard normal distribution. Since the normal distribution is symmetric, the P-value for the two-sided test is double this value. This two-sided P-value is equivalent to computing $\Pr(Z^2 > z_{Pe,obs}^2)$ where Z^2 is the square of a standard normal deviate, which has a chi-squared distribution with 1 degree of freedom. This is identical to the standard Pearson chi-squared test for a (2×2) table. Yates (1984) describes how Pearson originally thought the reference chi-squared distribution had three degrees of freedom rather than one, an error that was corrected by Fisher (1922b).

b. **Barnard's (1945) CSM (condition on a single margin) test (B)** also conditions on the (n_1, n_2) margin and computes the P-value as

$$\max_{0 \leq \pi \leq 1} \; p(Z_{Pe} \geq z_{Pe,obs} \mid \pi, H_0),$$

   where $\pi = \pi_1 = \pi_2$ is the overall probability that $B = 1$ under H_0. See also Suissa and Shuster (1989). This is a conservative test, in that it yields the largest P-value over all values of the nuisance parameter π. It is sometimes called the "exact" unconditional test, and it is included in StatXact software.

c. **Fisher exact test (F):** The Pearson chi-squared test is approximate in that it assumes large samples. An exact P-value is obtained using the Fisher exact test, which is based on repeated sampling from a reference distribution that conditions on both margins (n_1, n_2) and (m_1, m_2). From (2.1), the conditional distribution of x is:

$$\Pr(x \mid \pi_1, \pi_2, n_1, n_2, m_1)$$

$$= \frac{\binom{n_1}{x}\binom{n_2}{m_1 - x} \pi_1^x (1-\pi_1)^{n_1-x} \pi_2^{m_1-x}(1-\pi_2)^{n_2-m_1+x}}{\displaystyle\sum_{x'=1}^{\min(m_1, n_1)} \binom{n_1}{x'} \pi_1^{x'}(1-\pi_1)^{n_1-x'} \binom{n_2}{m_1-x'} \pi_2^{m_1-x'}(1-\pi_2)^{n_2-m_1+x'}}. \qquad (2.3)$$

Reparametrizing (π_1, π_2) into the odds ratio and odds product

$$\theta = \frac{\pi_1}{1-\pi_1} \Big/ \frac{\pi_2}{1-\pi_2}, \phi = \frac{\pi_1}{1-\pi_1} \times \frac{\pi_2}{1-\pi_2},$$

the terms in (2.3) involving the nuisance parameter ϕ cancel, and we obtain

$$\Pr(x \mid \theta, n_1, n_2, m_1) = \frac{\binom{n_1}{x}\binom{n_2}{m_1-x}\theta^x}{\displaystyle\sum_{x'=1}^{\min(m_1, n_1)} \binom{n_1}{x'}\binom{n_2}{m_1-x'}\theta^{x'}}.$$

Under H_0 the odds ratio $\theta = 1$, and this distribution reduces to the hypergeometric distribution

$$\Pr(x \mid \theta = 1, n_1, n_2, m_1) = \frac{\binom{n_1}{x}\binom{n_2}{m_1-x}}{\binom{n_1+n_2}{m_1}}.$$

The *P*-value for Fisher's exact test is then obtained by summing the hypergeometric distribution probabilities corresponding to values of x as extreme as or more extreme than that observed.

d. **Yates' continuity-corrected chi-squared test (Y):** The Fisher exact test calculation was difficult for moderate-sized tables without modern computers, and Yates (1934) introduced the continuity-corrected chi-squared test as an approximation of the Fisher exact test that was easier to compute. Yates' test compares

$$Z_Y^2 = \frac{\left(\left|x(n_2-y) - (n_1 - x)y\right| - N/2\right)^2}{m_1 m_2 n_1 n_2 / N},$$

with the standard normal distribution. Yates' showed that this test produced P-values much closer to the P-values for the Fisher exact test than for the Pearson test, particularly when the underlying hypergeometric distribution was symmetric.

The appropriate P-value for the two-sided Fisher exact test when the hypergeometric reference distribution is not symmetric is not obvious – how is "as or more extreme that the observed value" defined for the tail of the distribution not containing x, when the null distribution is asymmetric? A one-sided test compares Z_y (with the appropriate sign) with a standard normal distribution. Since this reference distribution is symmetric, the P-value for the two-sided test remains half the P-value for the one-sided test, and Yates (1984) recommends this choice for the Fisher exact test, too. This proposal receives quite a bit of push-back in the discussion of the paper, and how to define "extreme-ness" in the other tail is an issue with the P-value concept that does not seem to have an obvious single answer. For example, Cox (1977, Section 4.1) argues that "In the discrete case, we take the one-sided level q_0 plus the one- sided level from the other tail nearest to but not exceeding q_0 (Cox & Hinkley 1974, p. 79; Gibbons & Pratt 1975)."

Which of these four tests, P, B, F, or Y, is best? For large samples, they all give similar results, but the results can differ markedly in small or moderate samples. One distinguishing feature is that the hypergeometric reference distribution for F conditions on both margins, as does Y because it can be viewed as an approximation to F. The reference distributions for P and B condition on just one margin of the table, and hence include a larger set of possible tables. The rationale for conditioning on one margin rather than no margins is explained in the next section.

The F and Y tests tend to be more conservative than P and B, that is, yield larger P-values. Compare, for example, the P-values for P, F, and Y for the data in Table 2.2. Yates (1984) includes an extract from the simulations in Berkson (1978) to illustrate the conservatism of F and notes that the finding is replicated in other simulation studies.

There are two reasons for the relative conservatism of F and Y:

1. The test statistic for F and Y is "less extreme" than for P and B. A simple way to see this is that the continuity correction makes the chi-squared statistic for Y, the approximate version of F, smaller than the chi-squared statistic for P.

2. With discrete data, hypothesis tests for a fixed nominal size are inherently conservative. This conservatism is more pronounced for F and Y than for P and B.

TABLE 2.2

Example in Little (1989)

(a) Data					
		B		Total	Proportions
		1	2		
A	1	170	2	172	0.012
	2	162	9	171	0.056
Total		332	11	343	0.032

(b) *P*-values from alternative tests of H_0: $\pi_A = \pi_B$; H_a: $\pi_A > \pi_B$	
Test	**P-value**
Pearson Chi-squared (P)	0.016
Fisher exact (F)	0.030
Yates Continuity-Corrected Chi-squared (Y)	0.032

To elucidate point 2, for categorical data (as here), any test statistic has a discrete distribution under H_0, which means that the *P*-value can only take a finite set of possible values. A test with fixed nominal level then tends to be conservative, because the nominal rejection level must be less than the *P*-value. For example, if the closest possible *P*-value for test statistic less than or equal to 0.05 is 0.04, then the test cannot have a nominal size of exactly 0.05. This conservatism is more pronounced for F and Y than for P and B, because conditioning on both margins markedly reduces the set of possible tables and hence *P*-values – the null distribution is inherently coarser.

One approach to addressing this conservatism is to replace the *P*-value by the "mid-*P*," which Barnard attributes to Anscombe in his discussion in Yates (1984). The mid-*P*-value only includes half the probability of obtaining the observed value of the test statistic:

$$P_{\text{mid}} = 0.5 \times \Pr(T = t_{\text{obs}}) + \Pr(T > t_{\text{obs}} \mid H_0).$$

This suggestion has received considerable support – see, for example, Lyderson, Fagerland and Laake (2009) – but has not achieved widespread acceptance.

2.2.4 The Role of Ancillarity

Advocates of P and B – see, for example, Lyderson, Fagerland and Laake (2009) – argue that F and Y are too conservative and sacrifice power. Yates (1984), however, argues that the additional power of P and B is spurious and misleading, because the second margin is *approximately ancillary* for the parameter of interest θ, and hence should be fixed in repeated sampling. Thus, the question of whether the second margin should be fixed in repeated sampling become central to the argument.

For a discussion of ancillary statistics, see for example Ghosh, Reid and Fraser (2010). Suppose (x, a) is a minimal sufficient statistic for a parameter θ. Then, a is an *ancillary* statistic if its marginal distribution does not depend on θ, that is:

$$f(x, a \mid \theta) = f_A(a) f_{X|A}(x \mid a, \theta).$$

More generally, if (x, a) is a minimal sufficient statistic for parameters θ and ϕ, then a is S-ancillary for θ if

$$f(x, a \mid \theta, \phi) = f_A(a \mid \phi) f_{X|A}(x \mid a, \theta).$$

Fisher argued that the reference set for any test should condition on any ancillary statistics, whether or not these statistics are fixed by the sample design. Yates and many of the discussants in Yates (1984) take this position. The following three examples motivate this idea:

Example 2.1. Fisher's flowering plants

In his discussion of Yates (1984), Barnard describes the following example in a letter he received from Fisher. Suppose n plants are sampled, and a is the number of plants that germinate. The probability that a plant germinates is ϕ and is known to be 0.75. Plants that germinate can be red or purple, and $\theta = \Pr(\text{plant is purple})$. The question of interest is to test the null hypothesis $H_0: \theta = 0.5$ against the one-side alternative $H_a: \theta > 0.5$. Germination is independent of plant color, so a is an ancillary statistic for θ. Suppose $n = 4$ plants are sampled and all four germinate and are purple, so, letting y denote the number of purple plants, the data are $n = a = y = 4$.

The P-value conditioning on the number of plants that germinate is

$$\Pr(y = 4 \mid a = 4, n = 4) = 1/2^4 = 1/16.$$

The P-value averaging over the number of plants that germinate is

$$\Pr(y = 4, a = 4 \mid n = 4) = \Pr(y = 4 \mid a = 4, n = 4) \times \Pr(a = 4)$$

$$= (1/16)(3/4)^4 = (1/16)(81/256),$$

because all other combinations of y and a are less extreme. Thus, including the probability that all four plants germinate reduces the P-value by $81/256$; but to quote Fisher (as reported by Barnard):

> what if someone else had discovered how to get his plants to flower every time? He would, surely, justifiably complain if he, getting the same result, had it judged non-significant at 5 per cent, just because of his skill in horticulture?

Accordingly, Fisher argued that the *P*-value for the reference set that conditions on the number of measurements actually obtained – that is, the ancillary statistic *a* – is the appropriate one. Barnard "after long meditation" was forced to agree, leading him to abandon the CSM test for the (2×2) table problem. Fisher asserted that "Barnard is the only statistician who has ever admitted he was wrong."

Some statisticians think that Barnard should not have changed his mind! For example, Lyderson, Fagerland, and Laake (2009) recommend Barnard's CSM test, and the author of the current Wikipedia article on the CSM test (https://en.wikipedia.org/wiki/Barnard%27s_test) writes

> Under specious pressure from Fisher, Barnard retracted his test in a published paper, however many researchers prefer Barnard's exact test over Fisher's exact test for analyzing 2×2 contingency tables, since its statistics are more powerful for the vast majority of experimental designs, whereas Fisher's exact test statistics are conservative.

I wonder how the formidable Fisher might have reacted to that quote! I guess he might have said that it is the additional power of the unconditional tests that is specious, not his example. What do you think?

Example 2.2. Cox's measuring instruments

Cox (1958) presents the following example, which also argues for conditioning on ancillary statistics. Given that it is a famous example, I quote it in full (italics are mine):

> Statistical methods work by referring the observations to a sample space Σ of observations that might have been obtained… Suppose that we are interested in the mean θ of a normal population and that, by an objective randomization device, we draw either (i) with probability 1/2, one observation, x, from a normal population of mean θ and variance σ_1^2 or (ii) with probability 1/2, one observation x, from a normal population of mean θ and variance σ_2^2 where σ_1^2, σ_2^2 are known, $\sigma_1^2 \gg \sigma_2^2$, and where we know in any particular instance which population has been sampled…. The sample space formed by indefinite repetition of the experiment is clearly defined and consists of two real lines Σ_1, Σ_2, each having probability 1/2, and conditionally on Σ_i there is a normal distribution of mean θ and variance σ_i^2.
>
> Now suppose that we ask, accepting for the moment the conventional formulation, for a test of the null hypothesis $\theta = 0$, with size say 0.05, and with maximum power against the alternative θ', where $\theta' = \sigma_1 \gg \sigma_2$. Consider two tests. First, there is what we may call the conditional test, in which calculations of power and size are made conditionally within the particular distribution that is known to have been sampled. This leads to the critical regions $x > 1.64\sigma_1$ or $x > 1.64\sigma_2$, depending on which distribution has been sampled.
>
> This is not, however, the most powerful procedure over the whole sample space. An application of the Neyman-Pearson lemma shows

that the best test depends slightly on $\theta', \sigma_1, \sigma_2$, but is very nearly of the following form. Take as the critical region

$x > 1.28\sigma_1$, if the first population is sampled

$x > 5\sigma_2$, if the second population is sampled

Qualitatively, we can achieve almost complete discrimination between $\theta = 0$ and $\theta = \theta'$ when our observation is from Σ_2, and therefore, we can allow the error rate to rise to very nearly 10% under Σ_1. It is intuitively clear, and can easily be verified by calculation, that this increases the power, in the region of interest, as compared with the conditional test.

Now if the object is to make statements by a rule with certain specified long-run properties, the unconditional test just given is in order, *although it may be doubted whether the specification of desired properties is in this case very sensible. If, however, our object is to say "what we can learn from the data that we have," the unconditional test is surely no good.* Suppose that we know we have an observation from Σ_1. The unconditional test says that we can assign this a higher level of significance than we ordinarily do, because if we were to repeat the experiment, we might sample some quite different distribution. But this fact seems irrelevant to the interpretation of an observation, which we know came from a distribution with variance σ_1^2. That is, our calculations of power, etc., should be made conditionally within the distribution known to have been sampled, that is, if we are using tests of the conventional type, the conditional test should be chosen.

To sum up, *if we are to use statistical inferences of the conventional type, the sample space Σ must not be determined solely by considerations of power, or by what would happen if the experiment were repeated indefinitely.* If difficulties of the sort just explained are to be avoided, Σ should be taken to consist, so far as is possible, of observations similar to the observed set S, in all respects, which do not give a basis for discrimination between the possible values of the unknown parameter θ of interest. *Thus, in the example, information as to whether it was Σ_1 or Σ_2 that we sampled tells us nothing about θ, and hence, we make our inference conditionally on Σ_1 or Σ_2.*

Berger and Wolpert (1988, Section 2.1) assemble other examples that argue for conditioning on ancillary statistics on intuitive grounds. I discuss that monograph in more detail in Chapter 5 on the likelihood principle.

Pratt (1977) suggests that approaches to inference that do not condition on ancillaries fail the following adequacy principle

THE PRINCIPLE OF ADEQUACY. A concept of statistical evidence is (very) inadequate if it does not distinguish evidence of (very) different strengths.

Following the logic of Example 2.2 for the (2×2) table question, Cox writes in his discussion of Yates (1984):

> I accept three main theses of the paper, that the test should be conditional, that concentration on achieving preassigned magic levels like 0.05 rather than calculating P-values is misguided, and that by and large the power comparisons reported in the literature are irrelevant or worse.

Cox had a wry sense of humor and concludes his discussion with a joke. Playing on the military acronyms for non-commissioned officer (NCO), company sergeant major (CSM), and regiment sergeant major (RSM), he writes:

> On a point of terminology, perhaps the standard {CSM} test should be called the RSM test (remain with the same margins). It would thus take complete dominance not just over all other NCO's but over everyone else in sight.

My third example returns to inference for the (2×2) table:

**Example 2.3. Inference about cross-class proportions
in the "no margins fixed" design**

Returning to the (2×2) table problem, suppose Table 2.1 is obtained from the "no margins fixed design," where a sample of size N is randomly selected from the population, and Table 2.1 is the resulting cross-classification by two survey variables A and B, which then have a multinomial distribution with joint probabilities $\pi_{jk} = \Pr(A = j, B = k)$. Suppose the question of interest concerns the conditional distribution of B given A, specifically the probability that $B = 1$ in the subpopulations with $A = 1$ and $A = 2$:

$$\theta = (\pi_{1\cdot 1}, \pi_{1\cdot 2}), \quad \pi_{1\cdot j} = \Pr(B = 1 \mid A = j).$$

In survey sampling terminology A is called a *cross-class*, and the inferences concern cross-class proportions. In repeated sampling, the subsample size n_1 with $A = 1$ is Binomial with probability $\phi = \Pr(A = 1)$, and A is S-ancillary for the parameter of interest θ. Conditioning on A yields inferences that condition on the sample sizes n_1, n_2 in the two subpopulations. But for repeated samples with this sampling design, the sample sizes n_1, n_2 are not fixed but vary from sample to sample. In these repeated samples, the distribution of the usual estimate of $\pi_{1\cdot j}$, $\hat{\pi}_{1\cdot j} = p_j$, is technically not even well-defined, because with finite probability in repeated sampling, $n_j = 0$. Conditioning on the S-ancillary statistic and the cross-class sample sizes (n_1, n_2) seems clearly appropriate here.

These examples all suggest that repeated-sampling inferences should condition on ancillary statistics. This conflicts with the popular idea that which margins are fixed should be solely determined by the design. It seems clear

that margins that are fixed by design should also be conditioned in the reference set for inferences. Thus, F is the appropriate test for the "both margins
fixed" design in Section 2.2.1. Since the A-margin is fixed for the "one margin
fixed" design, conditioning on the A-margin is not controversial in this case,
and this is done in all tests in Section 2.2.3. However, a margin that is not
fixed by design may be ancillary, suggesting that it should be conditioned
in the reference set if the above arguments are accepted. In particular, the
A-margin in Example 2.3 is S-ancillary for the "no margins fixed" design
leading to Table 2.1, suggesting that that margin should be fixed in the reference set, even though it is not fixed by the design.

Should both margins be conditioned in the reference set (as in F) if none or
one margin is fixed in the design? The answer to that question is less clear.
In the "no margins fixed" and "one margin fixed" designs, the A-margin is
S-ancillary and hence should be fixed, but after conditioning on the A-margin,
the B-margin is not S-ancillary for the odds ratio θ; it has a distribution that
depends on both θ and the odds product ϕ. Yates (1984) argues that despite
this fact, the B margin should also be conditioned, because it is *approximately*
ancillary, in the sense that the information it contains about θ is small. But
approximate ancillarity is not clearly defined, so the question of whether to
condition on the second margin remains unresolved, as is evident from the
discussion of Yates' paper.

A radical way out of the dilemma is to adopt a Bayesian approach to the
problem, which avoids the need to choose a reference set in repeated sampling. This is the topic of the next section.

2.2.5 The Bayesian Approach

As discussed in Chapter 1, the likelihood for the parameters $\pi = (\pi_1, \pi_2)$ in the
"one margin fixed" design is the distribution in Eq. (2.1) treated as a function
of the parameters; constants not depending on the parameters can be omitted:

$$L(\pi \mid x, y, n_1, n_2) \propto \pi_1^x (1 - \pi_1)^{n_1 - x} \pi_2^y (1 - \pi_2)^{n_2 - y}. \tag{2.4}$$

Likelihood inference is based on the likelihood, and in particular <u>likeli-
hood ratios</u> for different values of the parameter

$$\mathrm{LR}(\pi, \pi' \mid x, y, n_1, n_2) = \frac{L(\pi \mid x, y, n_1, n_2)}{L(\pi' \mid x, y, n_1, n_2)}.$$

Repeated sampling plays no role once the model is set, so problems on
whether to condition on the margins do not arise – likelihood inference conditions on all the data. The most highly developed form of likelihood inference is
Bayesian inference, where the likelihood (2.4) is multiplied by a prior distribution $p(\pi)$ for the parameters, yielding (by Bayes rule) the posterior distribution:

$$p(\pi \mid x, y, n_1, n_2) \propto p(\pi) \times \pi_1^x (1 - \pi_1)^{n_1 - x} \pi_2^y (1 - \pi_2)^{n_2 - y}. \tag{2.5}$$

Statisticians, including Fisher (see Chapter 4), have attempted to achieve likelihood inference without a prior distribution, but Bayesian methods can be applied to any statistical model. The question becomes, how to choose the prior distribution; in a sense ambiguity about whether to condition on the margins is replaced by ambiguity in the posterior distribution obtained from different choices of prior.

The Bayesian answer corresponding to a one-sided test computes the posterior probability $\Pr(\theta > 1 \mid x, y, n_1, n_2) = 1 - \Pr(\theta < 1 \mid x, y, n_1, n_2)$, where θ is the odds ratio. The Bayesian answer corresponding to the two-sided test computes $\Pr(\theta = 1 \mid x, y, n_1, n_2)$, which is zero except for prior distributions that put a positive probability on $\theta = 1$. I say more about this later.

A common choice of weak prior assumes independent Jeffreys' priors for the proportions:

$$p(\pi) \propto \pi_1^{-1/2}(1 - \pi_1^{-1/2})\pi_2^{-1/2}(1 - \pi_2^{-1/2}), \tag{2.6}$$

which leads to independent Beta posterior distributions. For the data in Table 2.2, this choice leads to $\Pr(\pi_1 < \pi_2) = 0.013$, which is similar to the one-sided P-value for the Pearson test (0.016). In fact, Howard (1998) notes a mathematical relationship between these two estimates which means they are generally similar. Given that Bayes is "more conditional" than approaches that condition on both margins, one might expect the Bayesian answer to be closer to the Fisher or Yates P-values, but that is not the case.

This finding seems to provide some support for the unconditional frequentist tests, although the relationship between the P-value and Bayesian posterior probability is tenuous. Howard (1998), however, questions the assumption of prior independence of the probabilities, arguing for priors that allow for dependence between them. He illustrates the idea with the following example:

Example 2.4. Howard's (1998) Cattle Virus Example.

Howard considers a government committee that asks whether the proportion of cattle herds that are infected with a virus is higher in England or Scotland. If π_1 is the proportion for English herds and π_2 is the proportion for Scottish herds, then Howard asks whether the strength of evidence favors

$$H_1: \pi_1 > \pi_2 \text{ or } \bar{H}_1: \pi_1 < \pi_2$$

Howard notes that the probability that the proportion is exactly the same in the two populations is zero, so the null hypothesis is unnatural – a drawback of the frequentist approach is that the point null hypothesis $H_0: \pi_1 = \pi_2$ is often not sensible but is imposed so that the test statistic has a known distribution under the null hypothesis. For a forthright discussion of the problem with point null hypotheses, see Nester (1996). The Bayesian approach, in contrast, assigns posterior probabilities to H_1 and $\bar{H}_1$, treating the two hypotheses symmetrically.

Howard argues that independent priors for π_1 and π_2, as in Eq. (2.6), are not realistic:

> Suppose we were informed (before collecting any data) that the proportion of English cows infected was 0.8. With independent uniform priors we would now give H_1 a probability of 0.8 (because the chance that $\pi_2 > 0.8$ is still 0.2). In very many cases this would not be appropriate. Often we will believe (for example) that if π_1 is 80%, π_2 will be near 80% as well and will be almost equally likely to be larger or smaller.

Howard (1998) then discusses a range of prior distributions on π_1 and π_2 that allow for dependence between the probabilities, namely:

$$p(\pi) = e^{-(1/2)u^2}\,\pi_1^{\alpha-1}(1-\pi_1^{\beta-1})\pi_2^{\gamma-1}(1-\pi_2^{\delta-1}), \text{ where } u = \frac{1}{\sigma}\ln\left(\frac{\pi_1(1-\pi_2)}{\pi_2(1-\pi_1)}\right),$$

which extends the Haldane ($\alpha = \beta = \gamma = \delta = 0$), Jeffreys ($\alpha = \beta = \gamma = \delta = 1/2$) and Laplace ($\alpha = \beta = \gamma = \delta = 1$) priors to allow for prior dependence between the two probabilities. Table 2.3 (a combination of Howard's Table 4 and 5) shows the posterior distribution that $\pi_1 > \pi_2$ for a variety of data sets, and prior distributions with $\sigma = 1$. Allowing for dependent priors yields posterior probabilities closer to the *P*-values under F or Y than under P or B. Howard concludes:

> The use of the uncorrected Pearson chi-square test for the 2×2 table corresponds approximately to a Bayesian analysis using independent Jeffreys priors. Because the priors are independent, we feel this test is not sufficiently cautious (and so agree with Fisher's conclusion, although for a different reason). The recommended alternative would be to use dependent priors, as in Section 7, but for statisticians who wish to use a classical approach, the best option would seem to be the exact test. Even that may not be cautious enough for small sample sizes.

TABLE 2.3

Posterior Probabilities $\times\,100$ that $\pi_1 > \pi_2$ and Analogous *P*-values for Various Data Sets and Choices of Priors

Table (a, n_1-a, c, n_2-c)	Pearson	Fisher	Yates	Laplace Indep	Jeffreys Indep	Jeffreys $\sigma = 1$
(3,15,7,5)	0.9	2.4	2.4	1.1	0.9	3.2
(2,5,5,2)	5.4	14.3	14.3	6.6	5.3	13.9
(1,4,4,1)	2.9	10.3	10.3	4.0	2.7	12.3
(0,3,3,0)	0.7	5.0	5,1	1.4	0.4	10.4
(5,30,11,24)	4.4	7.7	7.7	4.7	4.3	6.9
(2,170,9,162)	1.6	3.0	3.2	1.7	1.3	3.2
(20,80,30,70)	5.1	7.1	7.1	5.2	5.1	6.0

Source: Table 4 in Howard (1998), reprinted with the permission of the Institute of Mathematical Statistics.

2.3 Conclusion

If you find the plethora of possible answers in Table 2.3 confusing, then welcome to statistics! Statistics is not mathematics, where answers are more cut
and dried!

If you dislike the ambiguity in the frequentist tests, you might dislike the
multiplicity of Bayesian answers even more. However, the Bayesian approach
avoids the issue of conditioning and yields answers that vary because of the
choice of prior, which is subject to debate as in Example 2.4. Both frequentists
and Bayesians might agree that answers tend to be non-unique and debat-
able when the sample size is small, so collecting more data may be the only
clear path to a definitive result.

In the next chapter, I look more generally at frequentist hypothesis testing and alternative approaches, featuring the rivalry between two statistical
giants, Fisher and Neyman. You can decide who you think wins that debate.

2.4 Some Thought Questions on This Chapter

1. The continuity-corrected chi-squared test is an approximation for
   which test, the Pearson chi-squared or the Fisher exact test?

2. What are the main arguments for using the Fisher exact test?

3. What are the main arguments for using the Pearson test?

4. Given two independent samples and a binary outcome, how many
   margins are ancillary for the odds ratio, zero, one, or two?

5. How persuaded are you by the arguments for conditioning on ancillary statistics in Section 2.2.4?

6. A common approach to the choice of test is to use Fisher's exact test
   for small sample sizes, and otherwise to use the Pearson test. An
   alternative approach is to use Fisher's exact test for small sample
   sizes, and otherwise to use the Yates' test. Why is that approach more
   internally consistent, and how many margins does it assume to be
   fixed?

7. What are the strengths and weaknesses of the Bayesian approach, as
   discussed in Howard (1998)?

8. Which side of the controversy do you come down on? (Noting that
   there is no "right" answer!)

3

Frequentist Flaps: Hypothesis Testing, Significance Testing, or Something Else?

The papers:

Fisher, R. A. (1955). Statistical methods and scientific induction. *J. Roy. Statist. Soc., Ser. B*, 17, 1, 69–78.

Neyman, J. & Pearson, E. S. (1933). On the problem of the most efficient tests of statistical hypotheses. *Phil. Trans. Roy. Soc. London, Ser. A*, 231, 289–337.

Neyman, J. (1956). Note on an article by Sir Ronald Fisher. *J. Roy. Statist. Soc., Ser. B*, 18, 2, 288–294.

Wasserstein, R. L. & Lazar, N. A. (2016). The ASA's statement on p-values: context, process, and purpose. *Amer. Statist.*, 70, 2, 129–133, with supplemental comments at: https://www.tandfonline.com/doi/full/10.1080/00031305.2016.1154108?cookieSet=1

Other readings:

Benjamin, D. J., Berger, J. O., et al. (2018). Redefine statistical significance. *Nat. Hum. Behav.*, 2, 1, 6–10.

Benjamini, Y., De Veaux, R., Efron, B., Evans, S., Glickman, M., Graubard, B. I., He, X., Meng, X.-L., Reid, N., Stigler, S. M., Vardeman, S. B., Wikle, C. K., Wright, T., Young, L. J., & Kafadar, K. (2021). ASA President's Task Force statement on statistical significance and replicability. *Harvard Data Science Review*, 3 (3). https://doi.org/10.1162/99608f92.f0ad0287

Lehmann, E. L. (1993). "The Fisher, Neyman-Pearson theories of testing hypotheses: One theory or two?" *J. Amer. Statist. Assoc.*, 88, 201–208.

McShane, B. B. & Gal, D. (2017). Statistical significance and the dichotomization of evidence, *J. Amer. Statist. Assoc.*, 112, 519, 885–895.

McShane, B. B., Gal, D., Gelman, A., Robert, C., & Tackett, J. L. (2019). Abandon statistical significance, *Amer. Stat.*, 73, suppl., 235–245.

3.1 Introduction

In this chapter, I discuss hypothesis testing, one of the main approaches for communicating evidence in statistics. I consider two controversies, one relatively narrow and the other much broader.

The narrow topic concerns the clash between Fisher and Neyman over approaches to hypothesis testing. Neyman's approach was laid out in his highly influential paper with Egon Pearson (Neyman & Pearson 1934), which many would say is one of the major contributions to mathematical statistics in the 20th century. The paper launched the Neyman-Pearson (NP) school of hypothesis testing, an approach to statistical inference that is still very much alive today. Central to this theory is the NP lemma, an important theorem which is still taught in classes on statistical theory. The paper defined some key concepts in hypothesis testing, such as Type 1 and Type 2 error, the critical region of a test under a choice of alternative hypotheses and efficient testing. The paper tied together previously disparate examples, creating a unified theory for single parameter problems that was expanded upon in future work.

The NP approach is decision-theoretic, and Fisher railed against the idea that scientific inference can be "reduced" to a decision-making problem. Fisher's (1955) paper is incendiary and controversial, and elicited an understandably icy response from Neyman (1956). Zabell (1992) described the exchange as "a battle which had a largely destructive effect on the statistical profession." The nature of the controversy is very clearly laid out in Lehmann (1993), which I have included in other readings. I'll rely on that paper quite a bit in my discussion of the topic.

The broader issue concerns the role of hypothesis testing in general and whether the deficiencies of the approach have contributed to the "replicability crisis" in science. This "crisis" has led some journals to avoid probabilistic expressions of statistical uncertainty altogether and prompted the American Statistical Association (ASA) to assemble committees of experts to consider the issue. The resulting guidance – not official ASA policy, as some have inferred – emerged in the form of the ASA's statement on P-values (Wasserstein & Lazar 2016.) The wide-ranging comments in the supplemental material, for the prosecution and the defense, are worthwhile reading. The case for hypothesis testing is presented in Benjamini et al. (2021), also including as other reading. Also in other reading, Benjamin et al. (2018) argue against the common 0.05 nominal value of significance tests, and McShane et al. (2017, 2019) argue more generally against the use of bright-line thresholds.

The ASA statement argues that hypothesis testing has a role, but there are better ways of communicating evidence, specifically confidence intervals and Bayesian inferences based on the posterior distribution. In Section 3.3, I discuss what I see as some of the major issues underlying the statement.

3.2 The Argument between Fisher and Neyman

3.2.1 Basics of a Frequentist Hypothesis Test

As discussed for the case of (2×2) tables in Chapter 2, the classical NP approach to hypothesis testing has the following basic elements:

A **null hypothesis** H_0 concerning the parameters of the model.

An **alternative hypothesis** H_a concerning values of the parameters when the null hypothesis is not true. A key idea in Neyman and Pearson (1933) was that an optimal test requires specification of this alternative hypothesis.

A **test statistic** T that has a known distribution if H_0 is true, and a "critical region" consisting of values of the test statistic representing departures from H_0.

The **P-value**, the probability under H_0 of observing values in the critical region as equal to or more extreme than the observed value t_{obs} of the test statistic:

$$P = \Pr(T \geq t_{obs} \mid H_0). \tag{3.1}$$

The **size** or **Type 1 error** of the test, α. In classical NP theory, the null hypothesis "rejected" in a test of size α if $P < \alpha$ and "accepted" (in better parlance, "not rejected") if $P > \alpha$. Common choices are $\alpha = 0.05$ or $\alpha = 0.01$.

The **Type 2 error** of the test, $\beta(\theta)$, where θ are model parameters, the probability of "accepting" (or better "not rejecting") the null hypothesis when the alternative hypothesis is true, that is:

$$\beta(\theta) = \Pr(T < t_{obs} \mid H_a \cdot \theta).$$

The **power** of the test is the probability of (correctly) rejecting the null hypothesis when the alternative hypothesis is true, that is:

$$\text{Power}(\theta) = 1 - \beta(\theta).$$

To this day, a standard frequentist **power analysis** determines the sample size such that, for a given nominal size (such as 5%), the test has at least a certain power (say 80%) of rejecting H_0 for an alternative value of the parameters θ considered to be of substantive importance.

Neyman and Pearson (1933) defined a uniformly most powerful (UMP) test as a test that has the highest power among all tests of fixed size α, for all values of the parameters. They then proved the NP Lemma, that for a point null hypothesis $H_0: \theta = \theta_0$ against a point alternative hypothesis $H_1: \theta = \theta_1$, the UMP test is based on the likelihood ratio (LR) test statistic

$$T = LR(\theta_0) = \frac{L(\theta_0 \mid \text{data})}{L(\theta_1 \mid \text{data})}. \tag{3.2}$$

For composite null hypotheses, where the null distribution of the test statistic involves unknown parameters ϕ, they propose maximizing the likelihood under the null and alternative hypotheses, yielding the LR test statistic:

$$T = LR(\theta_0) = \frac{L(\theta_0, \hat{\phi}(\theta_0) \mid \text{data})}{L(\hat{\theta}, \hat{\phi} \mid \text{data})}, \tag{3.3}$$

where $(\hat{\theta}, \hat{\phi})$ are the unrestricted maximum likelihood (ML) estimates of (θ, ϕ), and $\hat{\phi}(\theta_0)$ is the ML estimate of ϕ subject to $\theta = \theta_0$. This formulation ties together a number of previously disparate examples. While Fisher's (1922a) paper describes a general approach to estimation based on the likelihood, Eqs. (3.2) and (3.3) describe a starring role for likelihood in the hypothesis testing problem.

Fisher's approach to hypothesis testing specifies a null hypothesis, a test statistic that has a known distribution under the null hypothesis and measures departures from it, and a P-value given by Eq. (3.1). It does not include specific choices of alternative hypotheses, leaving the choice of test statistic to scientific considerations, and it does not involve a decision-theoretic "rejection" of the null hypothesis.

The main points of contention in the Fisher-Pearson argument concerned (a) inductive inference versus inductive behavior; (b) whether or not to specify an alternative hypothesis; (c) whether to reject or not reject the null hypothesis at a conventional choice of size; and (d) the choice of population, more specifically, the reference set for repeated-sampling calculations. I discuss each of these issues in the following subsections.

3.2.2 Inductive Inference versus Inductive Behavior

In the words of Lehmann (1993):

> Both Neyman and Fisher considered the distinction between "inductive behavior" and "inductive inference" to lie at the center of their disagreement. In fact, in writing retrospectively about the dispute... How strongly Fisher felt about this distinction is indicated by his statement in Fisher (1973, p. 7) that "there is something horrifying in the ideological movement represented by the doctrine that reasoning, properly speaking, cannot be applied to empirical data to lead to inferences valid in the real world."

Fisher's statement seems to me a serious exaggeration – I doubt that Neyman really embraced this "doctrine." Anyway, there is a basic distinction between "deductive" and "inductive" reasoning. Deduction consists of drawing logical conclusions from a set of assumptions, as in a mathematical proof. Induction consists of arguing from a particular set of facts – in statistics, the data – to a general theory. To use as an example the title of Naseem Taleb's (2007) book, "The Black Swan," the data might consist of the fact that all observed swans are white, and the inference might be that therefore "all swans are white" – a theory later disproved when black swans were "discovered" in the Antipodes.

Rather than trying to grasp the difference between "inductive inference" and "inductive behavior," terms not explicitly defined, I prefer to consider the specifics of the NP and Fisher approaches. The NP approach might be viewed as a probabilistic form of deduction – from the data, we behave as if either a null hypothesis is "accepted" or "rejected," and attempt to minimize the chance of making the wrong decision in repeated sampling. Fisher, who viewed himself primarily as a scientist rather than a mathematician, had a strong aversion to this conception of scientific inference – to him science progressed in the "Popperian" style where any worthwhile theory is capable of being tested empirically, and with the accumulation of evidence, it continues to stand up or eventually becomes discredited. No single experiment involves a decision; a decision evolves collectively over a sequence of experiments.

Fisher (1955) thought that probability was not up to the task of inductive inference – that a different concept, likelihood, was needed. He writes:

> Although some uncertain inferences can be rigorously expressed in terms of mathematical probability, it does not follow that mathematical probability is an adequate concept for the rigorous expression of uncertain inferences of every kind … More generally, however, a mathematical quantity of a different kind, which I have termed mathematical likelihood, appears to take …the place of probability as a measure of rational belief when we are reasoning from the sample to the population.

Fisher evolved the likelihood idea, where the model parameters are the arguments, into "fiducial inference," where "pivotal quantities" are used to justify probabilistic statements about parameters, without the need to specify a prior distribution. This is perhaps the one Fisher creation that has not been widely accepted, as I discuss in Chapter 4.

In a sense the dispute between Neyman and Fisher foreshadows the modern-day dispute between frequentist and Bayesian methods of statistical inference – NP standing for the frequentist approach, and Fisher for a quasi-Bayesian viewpoint (though he would have strongly disavowed the Bayesian characterization) Fisher's (1955) ruminations in the section entitled "Requirements of Inductive Inferences" strike me as having a Bayesian flavor, and he writes of Neyman:

> There seems here an entirely genuine inability to conceive that when new data are added in an inductive problem, previously correct conclusions are no longer correct.

Bayesian inference does allow the incorporation of information of prior studies formally into the inference through the prior distribution; I think a defect with classical hypothesis testing is the tendency to ignore previous studies, starting afresh with each new one.

At the heart of Neyman's probabilistic approach to inference was the notion of a confidence interval, which Fisher (1955) attacked:

> Neyman reinforces his choice of language by arguments much less defensible. He seems to claim that the statement (a) "θ has a probability of 5% of exceeding T" is a different statement from (b) "T has a probability of 5% of falling short of θ." Since language is meant to be used I believe it is essential that such statements, whether expressed in words or symbols, should be recognized as equivalent, even when θ is a parameter ... whilst T is directly calculable from the observations... we may point out that both statements are statements of the relationship in which T, or θ, stands to the other. Also, since probability is specified, the statements have meaning to a sufficiently well-defined population of pairs of these values. The statements do not imply that in this population of pairs of values either T or θ is constant, but also they do not exclude the possibility that one should be constant, and that variability should be confined to the other.

This statement seems to confuse what is being conditioned, as the statements $\Pr(\theta > T \mid \theta)$ and $\Pr(\theta > T \mid T)$ are not equivalent. As I tell my students, be careful what you condition on in probability statements. A similar cavalier attitude to probability statements underlies Fisher's arguments in support of fiducial probability.

When all is said and done, a large part of the dispute about inductive inference versus inductive behavior seems to me just semantics – both Neyman and Fisher were statisticians indulging in inductive inferences. Lehmann (1993) writes:

> Responding to earlier versions of these and related objections by Fisher to the NP formulation, Egon Pearson (1955, p. 206) admitted that the terms "acceptance" and "rejection" were perhaps unfortunately chosen, but of his joint work with Neyman he said that "from the start we shared Professor Fisher's view that in scientific inquiry, a statistical test is 'a means of learning'" and "I would agree that some of our wording may have been chosen inadequately, but I do not think that our position in some respects was or is so very different from that which Professor Fisher himself has now reached.... Professor Fisher's final criticism concerns the use of the term 'inductive behavior'; this is Professor Neyman's field rather than mine.

Lehmann also writes:

> There clearly are important differences, both in philosophy and in the treatment of specific problems. These were fiercely debated by Fisher and Neyman ... I believe that the ferocity of the rhetoric has created an exaggerated impression of irreconcilability.

3.2.3 Specifying an Alternative Hypothesis

Again, quoting Lehmann (1993):

> a central consideration of the NP theory is that one must specify not only the hypothesis H but also the alternatives against which it is to be tested. In terms of the alternatives, one can then define the type II error (false acceptance) and the power of the test (the rejection probability as a function of the alternative). This idea is now fairly generally accepted for its importance in assessing the chance of detecting an effect, ... determining the sample size required to raise this chance to an acceptable level, and providing a criterion on which to base the choice of an appropriate test... Fisher never wavered in his strong opposition to these ideas.... Fisher accepted the importance of [power] but denied the possibility of assessing it quantitatively.

Again, it seems that Fisher was opposed to the formalization of alternative hypotheses, preferring the choice of test to be left to scientific considerations in the given setting. It is true that one advantage of Fisherian significance testing is that it does not require the specification of alternatives, which might be unclear. For example, in Fisher's field of genetics, a null hypothesis of lack of linkage between a phenotype and a genotype can be tested without knowledge of the linkage mechanism. However, these days, just about any scientific grant that involves the collection of data needs a power analysis, such as that provided by the NP theory. There are other approaches, but the formal assessment of power as introduced by NP theory is viewed as essential.

3.2.4 The Introduction of a Fixed Nominal Level

Fisher did not approve of the NP theory rejecting or accepting (or better "failing to reject") the null hypothesis because it was a cog in the NP decision-theoretic wheel. Ironically, the conventional levels of size of 0.05 and 0.01 are attributed to Fisher, although these may have simply reflected the practical need to restrict tabulations of null distributions to a few key percentiles. Lehmann writes:

> Fisher... greatly reduced the needed tabulations by providing tables not of the distributions themselves but of selected quantiles... As Fisher wrote in explaining the use of his X^2 table "...In preparing this table we have borne in mind that in practice we do not want to know the exact value of P for any observed X2, but, in the first place, whether or not the observed value is open to suspicion. If P is between .1 and .9, there is certainly no reason to suspect the hypothesis tested. If it is below .02, it is strongly indicated that the hypothesis fails to account for the whole of the facts. We shall not often be astray if we draw a conventional line at .05 and consider that higher values of X2 indicate a real discrepancy."

3.2.5 Choice of Reference Set for Frequentist Calculations

In the section entitled "Repeated Sampling from the Same Population,"
Fisher (1955) raises the issue of the appropriate reference set for frequentist
calculations:

> The operative properties of an acceptance procedure, single or sequen-
> tial, are ascertained practically or conceptually by applying it to a series
> of successive similar samples from the same source of supply, and deter-
> mining the frequencies of the various possible results. It is doubtless in
> consequence of this that it has been thought, and frequently asserted,
> that the validity of a test of significance is to be judged in the same way.
> However, a rather large number of examples are now known in which
> this rule is seen to be misleading.

Fisher then provides as examples the question of whether to condition on
the covariates in a regression model, whether to condition on the second margin in a test of independence in the (2×2) table, as discussed in Chapter 2,
and the Behrens-Fisher problem of comparing means of two normal populations with differing variances, which I discuss in Chapter 4. Fisher advocated
conditioning on ancillary statistics, a practice with conflicts directly with the
NP idea of a UMP test; see, for example, Cox's measuring instrument example, Example 2.2.

Neyman and Pearson did not address this issue, and it seems to me one
aspect that Fisher got right. The issue does call into question the idea of a
UMP test, though it does not undermine the NP approach in general, since
there is no reason why the approach cannot be applied while conditioning
null sampling distributions on ancillaries.

Somewhat along the lines of Cox's discussion of Example 2.2, Lehmann
(1979) writes:

> Echoing Fisher, we might say that we prefer [the unconditional test] in an
> acceptance sampling situation where interest focuses not on the individual
> cases but on the long-run frequency of errors, but that we would prefer the
> [conditional test] in a scientific situation where long-run considerations are
> irrelevant and only the circumstances at hand (i.e., H or T) matter.

I'll have more to say about these two aspects of inference in the discussion
of calibrated Bayes inference in Chapter 6.

Some of my conclusions about this controversy:

1. Although there are undeniable differences between the NP and
   Fisherian approaches, they are perhaps inflated by Fisher's incendiary language.
2. I like Fisher's general approach to inference, which is more Bayesian
   than he might have allowed; Fisher's objections about conditioning

on ancillary statistics are substantive and are not to my mind well addressed by the NP theory.

3. Neyman seems to win on the utility of alternative hypotheses for power calculations, and on the nature of confidence intervals, contrasted with fiducial or Bayesian statements.

3.3 The *P*-value Debate

3.3.1 Introduction

> "Little p-value
> What are you trying to say
> Of significance?"
> ZILIAK'S HAIKU (IN SUPPLEMENT
> OF WASSERSTEIN ET AL. 2016)

Concerns about replicability of scientific findings have led statisticians to look at how statistics is used in applied research and what can be done to improve the situation. The ASA Board crafted a statement about one aspect of the question, the use of hypothesis testing in general and *P*-values in particular. The statement has been generally well received, and I reproduce the six main points below – the statement itself (Wasserstein & Lazar 2016) also includes commentary on each point.

1. "P-values can indicate how incompatible the data are with a specified statistical model.

2. P-values do not measure the probability that the studied hypothesis is true, or the probability that the data were produced by random chance alone.

3. Scientific conclusions and business or policy decisions should not be based only on whether a P-value passes a specific threshold.

4. Proper inference requires full reporting and transparency.

5. A P-value, or statistical significance, does not measure the size of an effect or the importance of a result.

6. By itself, a P-value does not provide a good measure of evidence regarding a model or hypothesis."

The extensive supplemental comments on the statement from a variety of perspectives are worthwhile reading. I summarize here what I view as some of the main points.

3.3.2 Some Important Points from the *P*-value Debate

1. **Nearly all statistical analysis involves untestable assumptions, and statistical measures such as *P*-values rely on the validity of those assumptions.**

   The results of hypothesis tests, as with other approaches to statistical inference, usually assume the validity of underlying assumptions, which are subject to misspecification. Some study designs provide more reliable results than other – in particular, in Part III, we discuss how randomized selection of units can avoid selection bias or allocation of treatments can avoid confounding of treatment effects. In analysis, an assumed statistical model can yield unreliable results if the underlying assumptions are violated. Defenders of hypothesis testing note that these issues are not confined to hypothesis testing.

2. **Requiring a small *P*-value can provide a useful shield from over-interpreting study findings that could be attributable to random variation. Small *P*-values can be a useful indicator of model misspecification, leading to better models or explanations of results.**

   For example, if a test of the null hypothesis that two treatments are equally effective yields a P-value of more than 0.10, then the evidence of a treatment difference is not compelling. This does not mean that the null hypothesis is true, only that the data are consistent with no drug difference. The use of hypothesis testing as a check against model misspecification is discussed further in Chapter 6. These tests do not necessarily require specification of an alternative hypothesis, such as the precise nature of a better statistical model.

3. **The logic of hypothesis testing is tortuous and often misunderstood, particularly by non-statisticians. The *P*-value is often wrongly interpreted as the probability that the null hypothesis is true and is a poor measure of evidence against the null hypothesis.**

   In teaching basic statistics to non-majors, I find that students are confused by the rather tortuous logic of hypothesis testing, as laid out in Section 3.2.1. The differing status of the null and alternative hypotheses is puzzling in settings where the two hypotheses should be viewed symmetrically, as in Example 2.4 above. It is not clear when the alternative is two-sided and when one-sided, when departures from the null are possible in both directions. McShane and Gal (2017) report on a study suggesting that even researchers who are primarily statisticians are prone to misuse and misinterpret *P*-values; in particular, they tend to interpret evidence dichotomously based on whether or not a *P*-value crosses the conventional 0.05 nominal level. This is poor practice, as described in point 6 below.

4. **The *P*-value is a poor measure of size of an effect because it confounds the effect size with the sample size.**
An important determinant of the *P*-value is the sample size, and we do not want estimates of the size of effects (as opposed to their precision) to depend on the size of the study. An effect size, however small, can result in a small *P*-value if the sample size is large enough. Thus, for "big data," *P*-values are nearly always small, even if the underlying estimates are not substantively large and may be vulnerable to bias from weak study design. This means that "statistical significance" needs to be carefully distinguished from "substantive significance" – a substantively minor effect can be statistically significant in a large study, and a substantively important effect may not be statistically significant in a small study. However, statistical and substantive significance are often confused.

A confidence or credible interval is a more intuitive and understandable tool for summarizing substantive significance. In a symmetric confidence interval, for example, the center of the interval is an estimate of the parameter of interest, and the width of the interval is an estimate of its uncertainty. Admittedly, the precise interpretation of confidence interval – a random interval that covers the true parameter value in a nominal fraction of repeated samples – is hard to follow and often misunderstood, and it is also subject to the choice of reference set that plagues classical hypothesis tests. However, a confidence interval often has a simpler interpretation as a (fixed) Bayesian credible interval, and I view the distinction between the confidence interval and the credible interval as more a mathematical nicety than a major issue.

Advocates of hypothesis tests argue that confidence intervals require a conventional choice of confidence coefficient, whereas *P*-values avoid the need to make this choice. Another point in their favor is that Fisherian significance tests do not require specifying an alternative hypothesis and choice of estimand. They can provide useful indicators of model misspecification or a guard against over-interpreting results that could be attributable to random variation. However, measurement is at the heart of statistics, and finding summary measures that capture the phenomenon of interest is often a profitable exercise.

5. **Classical hypothesis testing requires a test statistic with a known (or calculable) distribution under the null hypothesis. This often leads to point null hypotheses that are not scientifically useful.**
The important question is usually not whether differences are exactly zero – they rarely are – but rather whether the differences are substantively important. See, for example, Nester (1996).

6. **The conventional nominal level of 0.05 corresponds to weak evidence against the null hypothesis.**
   Comparisons with posterior odds ratios from Bayesian hypothesis testing suggest that the common nominal 0.05 significance level is weak evidence against the null, perhaps accounting for lack of replicability of results. See, for example, Johnson (2013), Benjamin et al. (2017), or the commentaries on Wasserman and Lazar (2016) by Johnson and Benjamin and Berger. The Bayesian posterior odds is the prior odds multiplied by the Bayes factor B, which integrates the likelihoods under the two hypotheses over the prior distribution. For a point null hypothesis,

$$B = \frac{\text{average likelihood of the observed data under } H_a}{\text{likelihood of the observed data under } H_0}.$$

   The Bayes factor B thus depends on the prior distribution, but an upper bound under quite general conditions is:

$$\bar{B} = -1/\left(eP\log(P)\right),$$

   where P is the P-value. Bayarri et al. (2016) propose $\bar{B}$ as a simple alternative to the P-value for measuring the influence of the data on the posterior odds. Benjamin and Berger (2019)) shows the relationship between this measure and the P-value. A P-value of 0.05 corresponds to very weak evidence against the null hypothesis ($\bar{B} = 2.44$), whereas a value of 0.005 corresponds to much more substantial evidence (($\bar{B} = 13.9$). This kind of evidence led Benjamin et al. (2017) to propose replacing $P = 0.05$ by $P = .005$ as a conventional nominal value. Hence, the limerick in my commentary (Little 2019, Val being Valen Johnson):

> "In statistics one thing do we cherish,
> P .05 we publish, else perish
> Val says that's so out-of-date, our studies don't replicate
> P .005, then null is rubbish!"

   McShane et al. (2019) criticize the proposal by Benjamin et al. (2017) because it embraces the idea of a cut-off in the first place, and as discussed in the next point, I don't think it is a good idea, but studying the relationship with Bayes factors is still useful for the interpretation of the P-value, even if no threshold is adopted.

7. **Bright-line thresholding should generally be avoided, unless the goal of an analysis is a decision. If the goal is a decision, then considerations other than the P-value are at least as important, such as the quality of the study design and data collection, and the validity of the resulting analysis.**

As noted in the Fisher and Neyman controversy, most modern statisticians do not advocate accepting or rejecting a null hypothesis based on whether the P-value is less than or more than a nominal size, such as 0.05. The practice of requiring for publication that a study shows demonstrates a statistically significant effect, at some nominal size such as 0.05, is pernicious and leads to poor statistical practices like "P-hacking," that is, testing multiple questions and tweaking statistical models to achieve statistical significance, and then selectively reporting results that achieve the nominal significance level. I discuss proposals to guard against this practice in Chapter 9.

Even with appropriate consideration of multiple comparisons, publishing only "statistically significant" findings distorts the evidence about a particular question. Studies with negative findings have information as well, and the main criteria for publication should be the importance of the question, the strength of the study design (see Part III), and the appropriateness of the statistical analysis.

These points have led some to argue for abolishing significance testing altogether, and simply presenting P-values as a measure of consistency with the null hypothesis. For example, Wasserstein, Schirm and Lazar (2019) write:

> The ASA Statement on P-Values and Statistical Significance stopped just short of recommending that declarations of "statistical significance" be abandoned. We take that step here. We conclude, based on our review of the articles in this special issue and the broader literature, that it is time to stop using the term "statistically significant" entirely. Nor should variants such as "significantly different," "p < 0.05," and "nonsignificant" survive, whether expressed in words, by asterisks in a table, or in some other way.

Defenders of the thresholding aspects of NP theory argue that sometimes a hard and fast decision is required. For example, the U.S. Food and Drug Administration needs to decide whether to approve a new drug, and drug companies seek clear guidance for when their new drugs meet the threshold for approval. P-values can play a role in screening out drugs where the statistical evidence is weak, but in practice, statistical significance is not the only (or even the most important) criterion. Other factors such as the trade-off between treatment benefits and side effects, the consistency of treatment effects across subpopulations, and the strength of the study design and analysis, should play a more important role.

3.3.3 Alternatives to Hypothesis Testing

If not hypothesis testing, then what? Avoiding all statements of statistical uncertainty, as adopted by some psychology journals, is throwing out the baby with the bathwater, because incorporating measures of statistical uncertainty is important when assessing study findings. Alternative measures of uncertainty seem to me preferable. Thus, Bayesian hypothesis tests answer directly the question of the probability of the truth of a hypothesis, albeit with additional assumptions concerning the form of a prior distribution on the parameters.

As indicated in the last section, I view confidence intervals and Bayesian credible intervals as preferable to hypothesis testing, at least if clear estimands can be identified. As my mentor David Wallace used to say at the University of Chicago, the heart of good statistics is good measurement, which means collecting data on meaningful measures using the best possible design and analysis.

3.4 Some Thought Questions on This Chapter

1. Summarize the main differences between the NP and Fisher approaches to hypothesis testing.
2. Fisher (1955) has three main three critiques of the NP school, concerning:
   i. "Repeated sampling from the same population,"
   ii. Errors of the "second kind,"
   iii. "Inductive behaviour," relating in particular to confidence interval statements.
      For each of these three, describe briefly Fisher's critique and Neyman's counterargument.
3. In the NP theory, why is it important to specify an alternative hypothesis for power calculations? What was Fisher's attitude to power analysis in statistical design?
4. Both confidence intervals and *P*-values are measures of statistical uncertainty and depend on the sample size. Why is it stated here that confidence intervals provide more useful measures of quantities of interest than confidence intervals? Do hypothesis tests have any advantages over confidence intervals?

4

Fiducial Inference and the Behrens-Fisher Problem

The papers:

Fisher, R. A. (1935b). The fiducial argument in statistical inference. *Ann. Eugenics*, 8, 391–398.

Ghosh, M. & Kim, Y.-H. (2001). The Behrens-Fisher problem revisited: a Bayes-frequentist synthesis. *Can. J. Statist.*, 29, 1, 5–17.

Welch, B. L. (1938). The significance of the difference between two means when the population variances are unequal. *Biometrika*, 29, 350–362.

Welch, B. L. (1956). Note on some criticisms made by Sir Ronald Fisher. *J. Roy. Statist. Soc. Ser. B*, 18, 2, 297–302.

Other readings:

Seidenfeld, T. (1992). R. A. Fisher's Fiducial argument and Bayes' Theorem. *Statist. Sci.*, 7, 3, 358–368.

Zabell, S. L. (1992). R.A. Fisher and the Fiducial argument. *Statist. Sci.* 7, 3, 369–387.

4.1 Introduction

Statistical ideas have often evolved in the context of apparently simple but iconic problems. In Chapter 3, I considered a basic statistics problem with a binary outcome, namely comparing proportions in a (2×2) table. In this chapter, I consider a parallel problem when the outcome is continuous and normal, namely comparing means of two independent normal samples with different means and variances. Like the analysis of the (2×2) table, this is a problem that is widely addressed in introductory statistics classes, for which the best solution is not obvious, and which raises fascinating general issues of statistical inference.

There is a correspondence between hypothesis testing and confidence intervals, with the $100(1 - \alpha)\%$ confidence interval including the set of null values of a parameter for which a test would not be rejected at the $100\alpha\%$ nominal level. In Chapter 3, I focused on hypothesis testing, so in this chapter, I'll focus more on interval estimation, while acknowledging that some

DOI: 10.1201/9781003395164-5

of the papers I describe approach the model from the hypothesis testing viewpoint.

As discussed in Chapter 3, confidence or Bayesian credible intervals are basic tools of statistical inference and are to my mind superior to hypothesis tests. The confidence interval was formally defined in an appendix of Neyman (1934), a paper which is primarily focused on survey sampling, and which I discuss in Chapter 13. The confidence interval is an interval computed from the data that includes the true parameter value in at least a given fraction $1 - \alpha$ of repeated samples. That is, $I_\alpha(\text{data})$ is a $100(1 - \alpha)\%$ confidence interval for a parameter θ_1 in a statistical model indexed by unknown parameters θ if

$$\Pr\left(\theta_1 \in I_\alpha(\text{data}) \mid \theta\right) \geq 1 - \alpha \text{ for all } \theta. \tag{4.1}$$

Here, $1 - \alpha$ is the confidence coefficient, chosen conventionally to be a fixed value such as 0.9, 0.95, or 0.99. The peculiar feature of the confidence interval is that it is a random interval that covers the true (fixed) parameter value in at least a certain percentage of repeated samples. Another feature is that Neyman did not require equality in Eq. (4.1), a reflection of the fact that exact confidence coverage is only achievable in a limited set of problems; in particular, it is not attainable for the Behrens-Fisher problem discussed here.

A Bayesian credible interval, on the other hand, is fixed given the data, and, under the assumptions of the Bayesian model and prior distribution, has posterior probability $1 - \alpha$ of covering the parameter θ_1, treated as a random variable, that is:

$$\Pr\left(\theta_1 \in I_\alpha(\text{data}) \mid \text{data}\right) = 1 - \alpha. \tag{4.2}$$

Converting frequentist statements like (4.1) about "data given parameters" to Bayesian statements like (4.2) about "parameters given data" is called the "inverse problem" in statistics. It is achieved via Bayes Theorem:

$$\Pr(\theta \mid \text{data}) = \Pr(\theta)\Pr(\text{data} \mid \theta) / \Pr(\text{data}), \tag{4.3}$$

where $\Pr(\theta)$ is the prior distribution of the parameters and $\Pr(\text{data})$ is a normalizing constant that does not depend on θ.

When possible, a good way of deriving confidence intervals like (4.1) is through a "pivotal quantity," which is a function of the parameters and data with a known distribution. Consider the following example.

Example 4.1. Inference for the difference of means of two independent normal samples when the ratio of the variances is assumed known

Suppose we have two independent normal samples:

$$(x_{i1}, i = 1, ..., n_1) \sim_{\text{iid}} G(\mu_1, \sigma_1^2), \quad (x_{i2}, i = 1, ..., n_2) \sim_{\text{iid}} G(\mu_2, \sigma_2^2); \quad \theta = (\mu_1, \sigma_1^2, \mu_2, \sigma_2^2),$$

where $G(\mu, \sigma^2)$ denotes the normal (Gaussian) distribution with mean μ and variance σ^2. The target for inference is the difference in means $\delta = \mu_2 - \mu_1$, and suppose $\sigma_2^2 = K\sigma_1^2$ where K is assumed known. The most common special case is $K = 1$, $\sigma_2^2 = \sigma_1^2$. The sample sizes (n_1, n_2), sample means $(\bar{x}_1, \bar{x}_2)$, and sample variances (s_1^2, s_2^2) in the two groups are sufficient statistics. A pivotal quantity for inference about δ is:

$$T = \frac{\bar{x}_2 - \bar{x}_1 - \delta}{s\sqrt{1/n_2 + K/n_2}}, \text{ where } s = \sqrt{((n_1 - 1)s_1^2 + (n_2 - 1)s_2^2/K)/(n_1 + n_2 - 2)},$$

because T has a Student's t distribution with $v = n_1 + n_2 - 2$ degrees of freedom for all θ. Thus:

$$\Pr\left(-t_{1-\alpha/2} < \frac{\bar{x}_2 - \bar{x}_1 - \delta}{s\sqrt{1/n_2 + K/n_2}} < t_{1-\alpha/2} \,\middle|\, \theta\right) = 1 - \alpha,$$

where $t_{1-\alpha/2,v}$ is the $100(1 - \alpha/2)$ percentile of the Student's t distribution with v degrees of freedom. Rearranging this expression,

$$\Pr(\delta \in I_{\text{pooled},\alpha} \,|\, \theta) = 1 - \alpha, \tag{4.4}$$

where $I_{\text{pooled},\alpha} = \bar{x}_2 - \bar{x}_1 \pm t_{1-\alpha/2,v} s\sqrt{1/n_2 + K/n_2}$. Thus, $I_{\text{pooled},\alpha}$ is an exact $100(1 - \alpha)\%$ confidence interval for δ.

Alternatively, suppose we adopt a Bayesian approach and assume the Jeffreys' prior

$$\pi(\mu_1, \mu_2, \sigma_1^2) \propto 1/\sigma_1^2.$$

for the unknown parameters. It can be shown that $I_{\text{pooled},\alpha}$ is also the Bayesian $100(1 - \alpha)\%$ posterior credible interval for δ, that is,

$$\Pr(\delta \in I_{\text{pooled},\alpha} \,|\, \text{data}) = 1 - \alpha. \tag{4.5}$$

Thus, $I_{\text{pooled},\alpha}$ has useful properties from both the frequentist and Bayesian perspectives.

Fisher called intervals like $I_{\text{pooled},\alpha}$ found by inverting pivotal quantities "fiducial intervals" and argued that they could be viewed as credible intervals that condition on the data. Loosely, he argued that:

$$\Pr(\text{data} \,|\, \theta) \Rightarrow \text{pivotal quantity} \Rightarrow_{\text{invert}} \Pr(\theta \,|\, \text{data}),$$

which Fisher claimed to have inferential validity as a credible interval, without the need to invoke a prior distribution.

For more detailed accounts of Fisher's fiducial approach, I refer to the articles in the other reading by Seidenfeld (1992) and Zabell (1992). I leave it to you to decide if the additional detail clarifies the logic. What is clear is that Fisher's argument has not achieved widespread acceptance – Zabell (1992) describes it as "Fisher's great failure." There are (at least) two problems with it. The first is that, as a matter of probability theory, the only way to transition rigorously from a confidence statement (conditional on data) to a credible statement (conditional on parameters) is via Bayes Theorem, which of necessity includes a prior distribution for the parameters. The second is that the pivotal quantities only exist for a limited class of problems. In fact, Fisher criticized Neyman's confidence interval definition as allowing inferences involving statistics that are not fully efficient (see Zabell 1992). Fisher attempted to generalize the fiducial argument to simultaneous inferences for more than one parameter, but these generalizations were suspect. Zabell (1992) writes:

> Where Fisher's 1930 paper had been cautious, careful and systematic, his 1935 paper was bold, clever but in many ways rash. For he now went on to conclude "In general, it appears that if statistics $T_1, T_2, T_3, \ldots$ contain jointly the whole of the information available respecting parameters $\theta_1, \theta_2, \theta_3, \ldots$, and if functions $t_1, t_2, t_3, \ldots$ of the T's and θ's can be found, the simultaneous distribution of which is independent of $\theta_1, \theta_2, \theta_3, \ldots$, then the fiducial distribution of $\theta_1, \theta_2, \theta_3, \ldots$ simultaneously may be found by substitution (Fisher 1935b, p395)." This sweeping claim illustrates the purely intuitive level at which Fisher was operating in this paper, and it was only towards the end of his life that Fisher began to express doubts about this position.

By "substitution," I think Fisher means acting as if the parameters $\{\theta_j\}$ have the distribution (data fixed) rather than the pivotal quantities $\{t_j\}$ having the distribution (parameters fixed). But the justification is not clear. In particular, I show in Section 4.2 that Fisher's application of this idea to the Behrens-Fisher problem yields an interval for the difference in means that does not have exact confidence coverage in repeated sampling.

Note that in this introduction, I have been careful to be clear about whether probability statements are conditioned on parameters or on data – something that Fisher did not always do. For example, Zabell (1992) quotes the following from Fisher (1945):

> A complementary doctrine of Neyman violating equally the principles of deductive logic is to accept a general symbolical statement such as
>
> $$\Pr(\bar{x} - ts < \mu < \bar{x} + ts) = 1 - \alpha, \tag{4.6}$$
>
> as rigorously demonstrated, and yet, when numerical values are available for the statistics $\bar{x}$ and s, so that on substitution of these and use of the 5 per cent value of t, the statement would read
>
> $$\Pr\{92.99 < \mu < 93.01\} = 95\%, \tag{4.7}$$

to deny to this numerical statement any validity. This is to deny the syllogistic process of making a substitution in the major premise of terms which the minor premise establishes as equivalent.

If Eq. (4.6) conditions on the parameters (μ, σ^2), it is a confidence interval statement concerning the distribution of $\bar{x}$ and s in repeated sampling. Conditioning on (μ, σ^2), (4.7) is meaningless because μ does not have a distribution. Eq. (4.7) makes sense as a Bayesian statement about the posterior distribution of μ given the data, but (4.6) and (4.7) are not equivalent statements because they condition on different events.

There is no obvious confidence interval for the difference in means when the ratio of the variances K is not (assumed) known – in fact, there is no exact confidence interval, in the sense of an interval that makes full use of the data and has exactly the nominal confidence coverage for all values of the underlying parameters. In Section 4.2, I consider three interval estimates of the differences in means:

a. the solution proposed by Welch (1938).
b. the solution proposed by Behrens (1929), which Fisher (1935b) endorsed as the fiducial interval.
c. the solution proposed by Ghosh and Kim (2001).

I reject the common wisdom concerning these solutions. Welch's solution is generally favored by frequentists but has flaws and is arguably inferior to the other solutions from the frequentist perspective. The Behrens-Fisher solution is Bayes for a conventional Jeffreys' prior, despite Fisher's antipathy to the Bayesian approach. Fisher justified it with his "fiducial" argument. Finally, the Ghosh and Kim procedure is less widely known and used, but in my view is superior to both the other two solutions. This solution can be viewed as an example of Calibrated Bayes, a philosophy that will be examined more closely in Chapter 6. It is remarkable what surprises are to be found in this most basic of statistics problems!

The three solutions considered here provide a clash between three distinct philosophies of inference, frequentist, Bayesian, and fiducial. Fisher introduced fiducial inference as a way of "making the Bayesian omelet without breaking the Bayesian egg," where "the egg" is the need to choose a prior distribution. As we saw in Chapter 1, Fisher rejected Bayes in favor of maximum likelihood, because the latter avoids the need to choose a prior distribution, but that method does not work so well in small samples. The papers by Seidenfeld (1992) and Zabell (1992) in the other readings provide useful insights into Fisher's thinking, which however remains obscure to many modern-day statisticians.

Rather than judging the alternative solutions by their etiology, it is better to judge them by their properties. In short, a good interval should (a) make efficient use of the available data, that is, should not (implicitly or explicitly) ignore or discard information; and (b) should have good properties under the assumed model, whether frequentist (in repeated sampling) or Bayesian (conditional on the observed data). When the variance ratio is assumed known, the interval $I_{\text{pooled},\alpha}$ has good frequentist and Bayesian properties. When the variance ratio is unknown, I argue in Section 4.3 that when viewed from this perspective, the Ghosh and Kim (2001) procedure is better than the alternatives.

4.2 Three Interval Estimates for the Difference in Means

4.2.1 Welch's Solution (W)

As discussed about, exact confidence intervals are generally based on pivotal quantities, which are functions of the data and parameters that have a known distribution for all values of the parameters. In particular, if the variances are known, the standard 95% confidence interval for δ is based on the pivotal quantity

$$Z = (\bar{x}_2 - \bar{x}_1 - \delta)/\sqrt{\sigma_1^2/n_1 + \sigma_2^2/n_2},$$

which has a standard normal distribution. In this section, I describe three interval estimates for δ when the variances and their ratio are not assumed known. Properties of these intervals are discussed in the Section 4.3.

The Welch (1934) solution replaces the unknown variances by sample variances, yielding the quantity

$$T = (\bar{x}_2 - \bar{x}_1 - \delta)/\sqrt{s_1^2/n_1 + s_2^2/n_2}. \tag{4.8}$$

Unlike the situation where the variances are assumed equal, T is not a pivotal quantity, in that it has a distribution that depends on the ratio of the unknown variances. Asymptotically, T has a standard normal distribution, and confidence intervals based on the normal distribution work fine if the sample sizes in both groups are sufficiently large, such as 30 or more.

The asymptotic approach does not work so well when sample sizes are small. The Welch solution approximates the unknown distribution by a Student's t distribution with degrees of freedom

$$v_{\text{Welch}} = \frac{\left(s_1^2/n_1 + s_2^2/n_2\right)^2}{\left(s_1^4/\left(n_1^2(n_1-1)\right) + s_2^4/n_2^2(n_2-1)\right)}, \tag{4.9}$$

which is based on a Satterthwaite (1946) approximation, equating the 4th moments to a t distribution. The Welch approximate degrees of freedom is always less than the pooled degrees of freedom, $n_1 + n_2 - 2$.

This approximate solution is widely used and can improve on the pooled variance interval in Eq. (4.4) if the variance ratio K deviates from its assumed value in that approach. It is simple to compute, and the approximation is generally good, yielding coverages close to the nominal level. However, it has theoretical flaws, as discussed in Section 4.3.

4.2.2 The Behrens-Fisher (BF) Solution

Although not the way Fisher would describe it, the Behrens-Fisher solution is most easily derived as the $100(1-\alpha)\%$ credible interval for the posterior distribution of $\delta = \mu_2 - \mu_1$ with a Jeffreys' prior on the parameters, namely

$$P(\mu_1, \mu_2, \sigma_1, \sigma_2) \propto \sigma_1^{-1}\sigma_2^{-1}. \tag{4.10}$$

From the Bayesian analysis of each group, the posterior distributions of μ_1, μ_2 are independent t centered at respective means $\bar{x}_1, \bar{x}_2$ with scales s_1, s_2 and degrees of freedom $n_1 - 1, n_2 - 1$. The posterior distribution of δ is simply the distribution of $\mu_2 - \mu_1$ derived from the joint posterior distribution of μ_1 and μ_2. A simple approach to computing the interval is simply to draw D means $\mu_1^{(d)}, \mu_2^{(d)}, d = 1, ..., D$ from their respective scaled t distributions, and setting $\delta^{(d)} = \mu_2^{(d)} - \mu_1^{(d)}$. Posterior credible intervals are then based on the set of respective draws $\{\delta^{(d)}, d = 1, ...D\}$.

Fisher derived the same interval using fiducial ideas. As discussed in Example 4.1, the fiducial distribution of δ when K is known is given by $\delta = \bar{x}_2 - \bar{x}_1 - Ts\sqrt{1/n_2 + K/n_2}$ where T is the pivotal quantity, which has a Student's t distribution with $n_1 + n_2 - 2$ degrees of freedom. The fiducial distribution of δ is obtained by integrating out K with respect to its fiducial distribution, which is based on s_1^2/s_2^2 having an F distribution centered at the variance ratio K.

4.2.3 The Ghosh and Kim (GK) Solution

This solution is the posterior credible interval for the Bayesian model with prior distribution

$$p(\mu_1, \mu_2, \sigma_1, \sigma_2) = \sigma_1^{-3}\sigma_2^{-3}(\sigma_1^2/n_1 + \sigma_2^2/n_2). \tag{4.11}$$

From a subjective Bayesian perspective, this prior distribution is a little odd because there seems no reason for the prior distribution for the variances depending on the sample sizes. The rationale for this choice of prior is discussed in the next section, which considers properties of all three proposed procedures.

4.3 Properties of the Methods

I showed in Example 4.1 that if the variances are assumed equal in the two samples, or more generally the ratio of the variances K is known, then an exact confidence interval for δ is available based on the t distribution. This interval is also the Bayesian posterior credible interval under the Jeffreys' prior. However, this inference is subject to bias if the assumption about the variance ratio is wrong, particularly if the samples sizes in the two groups are very different. Relaxing the assumption that K is known may yield better inferences. If sample sizes are large, then any of the approaches in Section 4.2 are preferable to the pooled variance method and perform similarly. The question addressed in this section is which of these methods work best in small or moderate sample sizes, when none of the procedures in Section 4.2 gives exact confidence coverage for all values of the parameters.

In Chapter 3, I discussed whether inferences should condition on ancillary statistics, as suggested by Fisher. Ghosh and Kim (2009) call the frequentist approach that conditions on ancillary statistics the "conditional frequentist" approach. For the Behrens-Fisher problem, the sample variances are S-ancillary (see Section 3.2.4,) in that their sampling distribution depends on the population variances but does not depend on the means (μ_1, μ_2), and hence the parameter of interest $\delta = \mu_2 - \mu_1$. Hence, the conditional frequentist coverage of confidence intervals for δ is of interest.

Fisher (1956) criticized Welch's procedure because he showed it was possible to obtain conditional confidence coverage, given the sample variances, that was always less than the nominal coverage, that is, was anti-conservative. He called this property "negatively-biased relevant subsets"; see also Buehler (1959) and Robinson (1976, 1982). Robinson (1982) showed that

$$\Pr(\,|T| > \alpha \,|\, s_1 = s_2) \geq \Pr(\,|t_{n_1+n_2-2}|\, > \alpha \text{ for all } \alpha > 0.$$

Thus, the set where $s_1 = s_2$ is a relevant subset where confidence intervals based on the Welch approximation cover the true value of δ less often than the nominal value suggests. This seems an undesirable property, if you believe in conditioning on ancillary statistics. An advantage of Bayesian credible intervals like BF and GK is that they do not have such negatively biased relevant subsets (Robinson 1976).

What are the properties of the BF and GK intervals with regard to confidence coverage? BF is Bayes with a Jeffreys' prior on the parameters. It always has conservative coverage, which is in accord with Neyman's definition (4.1), but at the expense of tending to be wider than the W or GK intervals.

Ghosh and Kim (2001) define a prior distribution π to be *probability matching to order u* if

$$\Pr(\delta \leq q_{(1-\alpha)} \,|\, \pi, \text{data}) = 1 - \alpha + o(N^{-u}),$$

TABLE 4.1

1000 Times the Proportion of Samples Where the True Value of δ is Below its 0.05 Posterior Quantile (Target = 50) and Below its 0.95 Posterior Quantile (Target = 950) when $\mu_1 = 2.0, \mu_2 = 0.0$ and $\sigma_2^2 = 1.0$

n_1,n_2	Prior	$\sigma_1^2 = 2.0$ 0.05	0.95	$\sigma_1^2 = 4.0$ 0.05	0.95	n_1,n_2	Prior	$\sigma_1^2 = 2.0$ 0.05	0.95	$\sigma_1^2 = 4.0$ 0.05	0.95
2,2	Jeffreys	7	991	10	989	2,3	Jeffreys	15	983	23	977
	Ghosh-Kim	17	979	24	979		Ghosh-Kim	36	965	42	952
20,2	Jeffreys	36	803	29	822	2,20	Jeffreys	36	788	43	788
	Ghosh-Kim	110	856	86	878		Ghosh-Kim	110	819	104	819
7,5	Jeffreys	31	969	34	959	5,7	Jeffreys	39	963	45	962
	Ghosh-Kim	41	955	45	948		Ghosh-Kim	49	952	52	952
15,10	Jeffreys	43	960	48	956	10,15	Jeffreys	44	952	44	949
	Ghosh-Kim	51	954	53	952		Ghosh-Kim	51	948	48	948
20,15	Jeffreys	46	958	44	955	15,20	Jeffreys	46	955	47	953
	Ghosh-Kim	52	953	49	951		Ghosh-Kim	53	951	49	952
30,20	Jeffreys	49	951	47	953	20,30	Jeffreys	48	953	48	952
	Ghosh-Kim	51	948	51	950		Ghosh-Kim	51	950	49	951

Source: Ghosh and Kim (2001) Tables 1 and 2. Reprinted with the permission of John Wiley & Sons Ltd. through PLSClear.

where $q_{(1-\alpha)}$ is the $(1-\alpha)$ quantile of the posterior distribution and N is the total sample size. The prior distribution (4.11) for the GK interval is designed so that the interval is second-order probability matching. In contrast, the BF interval is only first-order probability matching. This suggests that when viewed as a confidence interval, the GK interval tends to have closer to nominal coverage; in other words, it is *better calibrated*.

Ghosh and Kim (2001) demonstrate this property via a simulation study, where 1000 data sets are generated for various choices of sample sizes and parameters, and the proportion of samples where the true value of δ is below the 0.05 posterior quantile (target = 0.05) or below the 0.95 posterior quantile (target = 0.95) is determined, for the two choices of priors (4.10) and (4.11). Table 4.1 shows the results, taken from Tables 1 and 2 of 'Ghosh and Kim's (2001) paper. Except for the extreme cases with $(n_1,n_2) = (2,20)$ or $(20,2)$, the GK prior gives very well calibrated results, even when the sample sizes are quite small. The intervals based on the GK prior are clearly better calibrated than BF, which as noted tends to be conservative.

4.4 Summary

a. Inferential procedures like the intervals considered here should be assessed in terms of their properties, not their etiology.

b. W's confidence interval is flawed from the conditional frequentist perspective, in having negatively biased relevant subsets.

c. Fiducial inference is an attractive idea, but the justification seems questionable, and it is not a viable general approach. The fiducial argument has continued to confound statisticians, as discussed in Zabell's (1992) paper.

d. In particular, Fisher's fiducial justification of the BF interval seems to me suspect, and I find it more straightforward to view it as a Bayesian credible interval for the Jeffreys' prior, Eq. (4.10). It tends to give conservative coverage when viewed as a confidence interval.

e. The GK approach is Bayesian, but it is also attractive from a frequentist perspective, because viewed as a confidence interval it is well calibrated, even in small samples. Since it gives intervals that tend to be narrower and better calibrated than the BF interval, I think it is the clear winner in this comparison. It also has the useful property of being a Bayesian credible interval, though admittedly for a somewhat non-intuitive prior Eq. (4.11).

f. The approach of seeking Bayesian models leading to intervals with good confidence coverage is called Calibrated Bayes and is discussed in more generality in Chapter 6.

4.5 Some Thought Questions on This Chapter

1. What is a "pivotal quantity," and what role does it play in fiducial inference? Is there a pivotal quantity for (a) inference about a single mean from a normal sample, and (b) the B/F problem?

2. Summarize the properties of the Welch, BF and GK methods for computing interval estimates for δ in the Behrens-Fisher problem. Do you agree that the GK method is the best? If not, which method do you prefer and why?

3. I like to say "there are no exact frequentist solutions to the B/F problem; there are infinitely many Bayesian solutions, all exact but requiring a choice of prior." Explain.

4. Justify the statement "Bayesian posterior credible intervals should have good frequentist coverage, and frequentist confidence intervals are best if they can be interpreted as Bayesian posterior credible intervals."

5

Do You Like the Likelihood Principle?

The paper:

Birnbaum, A. (1962). On the foundations of statistical inference (with discussion). *J. Am. Statist. Assoc.*, 57, 269–326.

Other readings:

Berger, J. O. & Wolpert, R. L. (1988). The likelihood principle. *Inst Math Stat Lecture Notes-Monogr Series*, 6, 1–199.

Mayo, D. G. (2014). On the Birnbaum argument for the strong likelihood principle (with discussion). *Statist. Sci.*, 29, 2, 227–266.

5.1 Introduction

Birnbaum's paper argued that two principles that should be widely accepted, namely the conditionality principle (C) and the sufficiency principle (S), implied the likelihood principle (L), which states that, given the assumed truth of the statistical model, all the evidence in the data is contained in the likelihood function. That is,

$$C + S \Rightarrow L. \tag{5.1}$$

The paper stirred a lively controversy, because it suggests that frequentist approaches to statistical inference, based on repeated sampling properties of statistics under a statistical model, are inappropriate because they violate L. Thus, to be inferentially consistent, frequentists needed to disavow either C or S, both of which in Birnbaum's (1962) view are evidently true; in his later writings, Birnbaum had second thoughts about the validity of Eq. (5.1).

The stir that Birnbaum's (1962) paper created is illustrated by some quotes by prominent statisticians in the discussion:

> This paper is a landmark in statistics because it seems to me improbable that many people will be able to read this paper or to have heard it tonight without coming away with considerable respect for the likelihood principle... this paper is really momentous in the history of statistics. It would be hard to point to even a handful of comparable events. (L.J. Savage).

I haven't quite recovered from the shock of seeing that two principles I had thought reasonable and one which I had thought doubtful imply each other. (J. Cornfield).

The author claims that his principal result has "immediate radical consequences for everyday practice." However, the derivation has a gross blunder in elementary logic which immediately invalidates the result... The gimmick here is the innocent looking phrase "when adequate mathematical-statistical models can be assumed." What does this mean? (I. Bross).

I found the remarks of the last speaker [Bross] very stimulating because, although I pride myself on being a practical person, I have found the exact opposite of what was just said to be true... The fallacy which it is claimed exists in Birnbaum's paper is that he assumes the model to be correct. This objection could be made to any approach with precisely equal force. (G. Box).

Allan Birnbaum proves that $C + S \Rightarrow L$. Since modern Bayesians and other members of what George Barnard calls the "likelihood brotherhood" already accept (L), Birnbaum's result can win over only those statisticians who find (S) compelling for some other reason, such as by being compelled by the weight of Fisher's authority. Since I cannot see any other reason for finding (S) compelling, I am forced to the conclusion mentioned earlier that the paper is primarily a contribution to the sociology and not to the logic of statistics. (I.J. Good).

As Good's quote suggests, subjective Bayesians accept and follow the likelihood principle; they base inference on the posterior distribution obtained by multiplying the likelihood by the prior distribution, with repeated sampling not playing a role. Some "objective Bayes" methods, such as the use of Jeffreys' prior distributions that depend on the expected information, violate L, though arguably the violation is relatively minor in terms of its effect on the inference. In Section 5.2, I outline Birnbaum's argument, and in Section 5.3, I provide two canonical examples. In Section 5.4, I discuss the controversy, concluding that C seems to me reasonable, but the assumption of S in Eq. (5.1) is a weakness. In other reading, the monograph by Berger and Wolpert (1988) is a tour-de-force that generally favors L. The paper by Mayo (2014) provides a perspective more critical of L. The discussion of that paper suggests that the controversy persisted 50 years after Birnbaum's paper, and it no doubt still persists today.

5.2 Birnbaum's "Proof" of the Likelihood Principle

Let E_1 and E_2 denote two experiments, and E_j yields data x_j, which are assumed to follow a statistical model with density or mass function $f_j(x_j \mid \theta)$ indexed by a finite-dimensional parameter θ, which has the same interpretation in both experiments. Let $\tilde{x}_j$ denote the realized value of x_j. Define realizations of these experiments, $(E_1, \tilde{x}_1)$ and $(E_2, \tilde{x}_2)$, to be a **strong likelihood pair (SLP)** if the likelihoods of θ under the respective models are proportional. That is, for some function $k(x_1, x_2)$:

$$f_1(\tilde{x}_1, \theta)/f_2(\tilde{x}_2, \theta) = k(\tilde{x}_1, \tilde{x}_2) \text{ for all } \theta. \tag{5.2}$$

Let $\text{Inf}(\theta \mid E_j, \tilde{x}_j)$ denote inference about the parameter θ based on experiment E_j and realized data $\tilde{x}_j$. By "Inf," I mean a statistical inference such as a confidence interval for a component of θ, a hypothesis test of a particular set of values of θ, or a Bayesian posterior credible interval for a component of θ. The likelihood principle is then:

Likelihood Principle (L): if $(E_1, \tilde{x}_1)$ and $(E_2, \tilde{x}_2)$ are a SLP, then $\text{Inf}(\theta \mid E_1, \tilde{x}_1) = \text{Inf}(\theta \mid E_2, \tilde{x}_2)$.

That is, SLPs should yield the same statistical inferences.

Birnbaum (1962) links the experiments E_1 and E_2 via the following mixture experiment E_B.

E_B: let U be a random number between 0 and 1, and let $Z = 1$ if $U < 0.5$, $Z = 2$ if $U > 0.5$. If $Z = j$, perform experiment E_j, yielding data x_j.

Here, the subscript B denotes Birnbaum; the term "Birnbaumization" of E_1 and E_2 has been coined for this mixture experiment (Mayo 2014). We now define:

Weak Conditionality Principle (C):

$$\text{Inf}(\theta \mid E_B, \tilde{z} = j, \tilde{x}_j) = \text{Inf}(\theta \mid E_j, \tilde{x}_j) \tag{5.3}$$

In words, the inference about θ from performing E_B and obtaining $(\tilde{z} = j, E_j, \tilde{x}_j)$ is the same as the inference about θ from performing experiment E_j directly and obtaining $\tilde{x}_j$, without the preliminary randomization.

A sufficient statistic $t(x)$ for an experiment yielding data x is a statistic such that the densities

$$f_X(x \mid \theta) = f_T(t \mid \theta) \times f_{X\mid T}(x \mid t), \tag{5.4}$$

where the conditional distribution of x given t does not depend on θ. (Note that Eq. (5.4) differs from the definition of an ancillary statistic a, where $f_X(x \mid \theta) = f_A(a) \times f_{X\mid A}(x \mid a, \theta)$).

The Sufficiency Principle states that, assuming correctness of the underlying model, two outcomes x and x' of an experiment that yield the same value of a sufficient statistic should yield the same inferences. That is:

Sufficiency Principle (S). Let $t(x)$ be a sufficient statistic for the unknown parameter θ in an experiment E. Let x and x' be two possible outcomes of an experiment E. If $t(x) = t(x')$, then $\text{Inf}(\theta \mid E, x) = \text{Inf}(\theta \mid E, x')$.

Birnbaum (1962) argues that S implies that two outcomes from a particular experiment that yield the same likelihood should yield the same inferences –

he uses the more general term "evidence" (Ev) rather than my more specific term "inference." To quote from the paper:

> It has been shown in the general theory of sufficient statistics ... that if two outcomes x, x' of *one* experiment E determine the same likelihood function (that is, if for some positive c we have $f(x,\theta) = cf(x',\theta)$ for all θ, then there exists a (minimal) sufficient statistic t such that $t(x) = t(x')$. (In the case of any discrete sample space, the proof is elementary.) This, together with S, immediately implies
>
> **Lemma 1.** If two outcomes x, x' of any experiment E determine the same likelihood function, then they have the same evidential meaning: $Ev(E,x) = Ev(E,x')$."

Birnbaum then argues that if $(E_1, \tilde{x}_1)$ and $(E_2, \tilde{x}_2)$ are a SLP, the corresponding outcomes from the mixture experiment E_B, $(z = 1, \tilde{x}_1)$ and $(z = 2, \tilde{x}_2)$ lead to the same likelihood, so that:

$$\text{Under S: } Inf(\theta \mid E_B, z = 1, \tilde{x}_1) = Inf(\theta \mid E_B, z = 2, \tilde{x}_2). \tag{5.5}$$

Thus, combining (5.3) and (5.5):

Assuming C and S : $Inf(\theta \mid E_B, z = 1, \tilde{x}_1) = Inf(\theta \mid E_1, \tilde{x}_1) = Inf(\theta \mid E_2, \tilde{x}_2)$
$$= Inf(\theta \mid E_B, z = 2, \tilde{x}_2),$$

the central equality being the likelihood principle.

Birnbaum's argument does not spell out the sufficient statistic involved in S given outcomes of the mixture experiment $(E_1, \tilde{x}_1)$ and $(E_2, \tilde{x}_2)$. Berger and Wolpert (1988) define such a statistic:

$$T(Z = j, \tilde{x}_j) = \begin{cases} \tilde{x}_1, & \text{if } Z = 1; \\ \tilde{x}_1(\tilde{x}_2), & \text{if } Z = 2, \end{cases}$$

where $\tilde{x}_1(\tilde{x}_2)$ is the value of x_1 such that $(E_1, \tilde{x}_1)$, and $(E_2, \tilde{x}_2)$ are a SLP. Then,

$$f_j(\tilde{x}_j, \theta) = c_j(\tilde{x}_j) \times L(\theta),$$

where $L(\theta)$ is the common likelihood function for the SLP $(E_1, \tilde{x}_1)$ and $(E_2, \tilde{x}_2)$. Hence, T is a sufficient statistic, because

$$\Pr(Z = 1, x_1 = \tilde{x}_1 \mid T = \tilde{x}_1, \theta) = \frac{0.5 \times f_1(\tilde{x}_1, \theta)}{0.5 \times f_1(\tilde{x}_1, \theta) + 0.5 \times f_2(\tilde{x}_2, \theta)}$$
$$= \frac{0.5 \times c_1(\tilde{x}_1) \times L(\theta)}{0.5 \times c_1(\tilde{x}_1) \times L(\theta) + 0.5 \times c_2(\tilde{x}_2) \times L(\theta)},$$
$$= \frac{c_1(\tilde{x}_1)}{c_1(\tilde{x}_1) + c_2(\tilde{x}_2)},$$

which does not involve θ. This argument seems to assume a unique SLP mapping corresponding to each value of x_1 and x_2 in the two component models, because otherwise $\tilde{x}_1(\tilde{x}_2)$ does not have a clear definition.

I now consider two iconic and widely cited examples.

5.3 Two Examples

Example 5.1. Binomial versus Negative Binomial sampling

Suppose (E_1, x_1, n_1) denotes a binomial experiment, with outcome the number of successes x_1 in a fixed number of independent trials, n_1. Let $y_1 = n_1 - x_1$ denote the number of failures. If θ is the probability of a success, then the distribution of x_1 is binomial, with probability mass function

$$f_1(x_1 \mid n_1, \theta) = \binom{n_1}{x_1} \theta^{x_1}(1-\theta)^{y_1}, \text{ where } \binom{n_1}{x_1} = \frac{n_1!}{x_1!(n_1-x_1)!}. \quad (5.6)$$

Suppose (E_2, x_2, y_2) denotes a negative binomial experiment, where trials are carried out until a fixed number of failures y_2 are obtained, and the outcome is the number of successes x_2. If θ is the probability of a success, then the distribution of x_2 is then negative binomial, with probability mass function

$$f_2(x_2 \mid y_2, \theta) = \binom{x_2 + y_2 - 1}{x_2} \theta^{x_2}(1-\theta)^{y_2}. \quad (5.7)$$

Comparing (5.6) and (5.7), clearly $(E_1, \tilde{x}_1, \tilde{n}_1)$ and $(E_2, \tilde{x}_2, \tilde{y}_2)$ such that $\tilde{x}_1 = \tilde{x}_2 = \tilde{x}$ and $\tilde{n}_1 - \tilde{x}_1 = \tilde{y}_2 = \tilde{y}$ are a SLP. Define $t = (x, y)$. The conditional distribution of Z given T and θ has odds:

$$\frac{\Pr(z=1 \mid \tilde{x}, \tilde{y}, \theta)}{\Pr(z=2 \mid \tilde{x}, \tilde{y}, \theta)} = \left[\binom{\tilde{x}+\tilde{y}}{\tilde{x}}\theta^{\tilde{x}}(1-\theta)^{\tilde{y}}\right] / \left[\binom{\tilde{x}+\tilde{y}-1}{\tilde{x}}\theta^{\tilde{x}}(1-\theta)^{\tilde{y}}\right] = \frac{\tilde{x}+\tilde{y}}{\tilde{y}}, \quad (5.8)$$

which does not involve θ. So $f(Z,T\mid\theta) = F(T\mid\theta)f(Z\mid T)$, and T is a sufficient statistic.

L implies that for the SLP $(E_1, \tilde{x}, \tilde{x}+\tilde{y})$ and $(E_2, \tilde{x}, \tilde{y})$, inferences for E_1 and E_2 should be the same. But because x has a different distribution under the two experiments, frequentist inferences such as tests of null values of θ or confidence intervals for θ are different, conflicting with L.

This set-up is also an example where the "objective Bayes" analysis based on the Jeffreys' prior for θ violates L. If $L(\theta)$ denotes the likelihood, the Jeffreys' prior distribution is $\pi(\theta) \propto [I(\theta)]^{1/2}$, where $I(\theta)$ is the expected information $I(\theta) = E[\partial^2 \log L(\theta)/\partial\theta^2]$. For the binomial and negative binomial models, the Jeffreys' priors are $\pi(\theta) \propto \theta^{-1/2}(1-\theta)^{-1/2}$ and $\pi(\theta) \propto \theta^{-1}(1-\theta)^{-1/2}$, respectively. Use of these priors yields slightly different posterior distributions of θ for binomial and negative binomial experiments with the same likelihood.

Example 5.2. Data-Dependent Stopping Rules

An oft-cited test case for L concerns inference from fixed as opposed to sequential sampling. Suppose E_1 consists of taking a random sample of fixed size n, say $x = (x_1, ..., x_n)$, where $x_i \sim_{\text{ind}} G(\theta, \sigma^2)$, a normal (Gaussian) distribution with unknown mean θ and known variance σ^2. E_2 consists of sampling $y = (y_1, ..., y_m)$, where $y_i \sim_{\text{ind}} G(\theta, \sigma^2)$, θ and σ^2 are as for E_1, and the sampling rule is to sample until $\bar{y}_m > 1.96\sigma/\sqrt{m}$, where $\bar{y}_m = \sum_{i=1}^{m} y_i/m$. A SLP is then $(\bar{x}_n, n)$ and $(\bar{y}_m, m)$ where $m = n$, $\bar{x}_n = \bar{y}_m$. Under E_1, the nominal size for a test of $H_0: \theta = 0$ against the alternative $H_a: \theta \neq 0$ that rejects H_0 when $\bar{x}_n > 1.96\sigma/\sqrt{n}$ is $\alpha = 0.05$. Under E_2, the nominal level for this test is $\alpha = 1$, because the stopping rule guarantees rejection of H_0 regardless of the value of θ, and in particular when θ equals the null value $\theta = 0$. Thus, the results of this hypothesis test clearly differ for a SLP, contradicting L, which implies that the inferences should be the same.

Berger and Wolpert (1962) call E_2 "stopping when the data looks good." The frequentist hypothesis test differs for E_1 and E_2, and frequentists, who reject L, believe that this is appropriate – the stopping rule for E_2 is part of the evidence. Bayesians accepting L argue that the problem is that the frequentist hypothesis test is not the appropriate way to summarize the evidence. Specifically, framing the problem in terms of a point null hypothesis, without reference to the estimated size of effect, has the problems of a point null noted in Chapter 3; a Bayesian credible interval for θ from E_2, equivalent to the frequentist confidence interval, would always exclude zero, but in a large enough sample would also tend to preclude substantively important values if the true value of θ is small.

If the stopping rule in E_2 seems extreme and unwise, more realistic data-dependent stopping rules, such as in clinical trials with interim analyses, raise similar issues. The following example in Berger and Wolpert (1988) illustrates the thorny issues that can arise in a frequentist analysis (in this quote N denotes the normal distribution):

> A scientist enters the statistician's office with 100 observations, assumed to be independent and from a $N(\theta, 1)$ distribution. The scientist wants to test $H_0: \theta = 0$ versus $H_1: \theta \neq 0$. The current average is $\bar{x}_n = 0.2$, so the standardized test statistic is $z = \sqrt{n}|\bar{x}_n - 0| = 2$. A careless classical statistician might simply conclude that there is significant evidence against H_0 at the 0.05 level. But a more careful one will ask the scientist, "Why did you cease experimentation after 100 observations?" If the scientist replies, "I just decided to take a batch of 100 observations," there would seem to be no problem, and very few classical statisticians would pursue the issue. But there is another important question that should be asked (from the classical perspective), namely: "What would you have done had the first 100 observations not yielded significance?" To see the reasons for this question, suppose the scientist replies: "I would then have taken

another batch of 100 observations." This reply does not completely specify a stopping rule, but the scientist might agree that he was implicitly considering a procedure of the form: (a) take 100 observations; (b) if $|\sqrt{100}\bar{x}_{100}| \geq k$ then stop and reject H_0; (c) if $|\sqrt{100}\bar{x}_{100}| < k$ then take another 100 observations and reject if $|\sqrt{200}\bar{x}_{200}| \geq k$. For this procedure to have level $\alpha = 0.05$, k must be chosen to be 2.18 (Pocock 1977). Since the actual data had $|\sqrt{100}\bar{x}_{100}| = 2 < 2.18$, the scientist could not actually conclude significance, and hence would have to take the next 100 observations. This strikes many people as peculiar. The interpretation of the results of an experiment depends not only on the data obtained and the way it was obtained, but also upon *thoughts* of the experimenter concerning plans for the future.

Berger and Wolpert go on to speculate that the results after the next 100 observations are still not quite significant, and whether to proceed further depends on whether a pending grant renewal is funded. So, the appropriate sequential frequentist analysis depends on the outcome of that event as well. They conclude:

> Note that we are not faulting the classical statistician here for ascertaining and incorporating the stopping rule in the analysis. If one insists on utilization of frequentist measures, such involvement of the stopping rule (even if it exists only in the imagination of the experimenter) is mandatory. The need here for involvement of the stopping rule clearly calls the basic frequentist premise into question, however.

Frequentists might respond that the dependence of inference on the stopping rule is a property of frequentist inference, not a weakness. In clinical trial practice, the problems raised by this example are ameliorated by specifying an explicit plan for interim analyses in the study protocol.

From the calibrated Bayes perspective discussed in Chapter 6, there is a disadvantage of data-dependent stopping rules, namely that they are less well calibrated and more sensitive to specification of the prior distribution than inferences based on a fixed sample size (Rosenbaum & Rubin 1984).

5.4 Discussion

Statisticians rejecting L either find flaws in Birnbaum's "proof" or reject or modify C or S. I discuss each of these angles in the next three subsections.

5.4.1 Challenges to the Birnbaum Proof

Mayo (2014) argues that there are flaws in Birnbaum's argument, a contention that is supported by some discussants of her paper, though sometimes for different reasons, and rejected by others, notably Dawid (2014). The perplexing nature of Birnbaum's "proof" is that knowledgeable experts come to such radically different conclusions, and I don't blame you if you come away from Mayo's paper and the ensuing discussions feeling even more confused than when you started. In this section, I outline some of the challenges to Birnbaum's result, while admitting that I find some of the finer points of disputed logic elusive.

5.4.1.1 Measure-theoretic Issues

A technical issue is that Birnbaum's proof is relatively straightforward for discrete data but requires some measure-theoretic elaborations to handle continuous data. Difficulties in Birnbaum's argument arise for continuous distributions where the probability of particular values of x are zero, and in nonparametric models (e.g. Robins & Wasserman, 2000). Berger and Wolpert (1988, Section 3.4) discuss these difficulties and propose resolutions.

5.4.1.2 Murky Logic in Birnbaum's Proof

In Section 4.2 of Birnbaum (1962), he writes:

> It can be shown that (S) is implied mathematically by (C). (The method of proof is the device of interpreting the conditional distribution of x, given $t(x) = t$, as a distribution $G(h)$ defining a mixture experiment equivalent to the given experiment.) This relation will not be discussed further here, since there seems to be little question as to the appropriateness of (S) in any case.

If C and S implies L, and C implies S, as suggested by this quote, then logically C implies L, so why is S invoked as a condition? Thus, Kempthorne writes in his discussion of Birnbaum (1962):

> I found Dr. Birnbaum's paper somewhat obscure as regards the relationships among S, C and L. He states (1) "S is implied mathematically by C"; (2) Lemma 1: L implies S; (3) Lemma 2: L implies and is implied by S and C. In particular, because C implies S, lemma 2 should read in part, it seems, C implies L.

This raises the important question of whether C by itself implies L, so that S is superfluous. Birnbaum's sketch proof that "S is implied by C" invokes a stronger version of Conditionality than C, and I find it hard to follow. The idea that a version of C implies L also recurs in Evans, Fraser and Monette (1986). However, I have difficulty understanding how the equivalence in Eq. (5.5) is established without invoking a sufficiency condition.

5.4.1.3 Are S and C Contradictory?

The indicator for which component in the Birnbaum mixture experiment is conducted is not part of the sufficient statistic invoked under S; see, for example, the sufficient statistic (x, y) in Example 5.1. But the indicator is the ancillary statistic that is conditioned under C, and as such seems an important part of the evidence. Thus, Evans (2014) states in his discussion of Mayo (2014) (italics are mine):

> The relevance of frequentism to Mayo's paper lies in the author's position that Birnbaum's argument is basically a violation of the principle. An argument is provided for why the joint application of S and C used in Birnbaum's proof constitute such a violation. I accept Mayo's reasoning. In fact, I think it is somewhat similar to the argument put forward in Evans, Fraser & Monette (1986) that the applications of S and C in the proof are incorrect because S *discards as irrelevant precisely the information used by* C *to form the conditional model.* So the justifications for S and C contradict one another in the proof and this doesn't seem right.

5.4.1.4 Is Birnbaumization a Valid Experiment?

Some have argued that Birnbaum's argument only applies if the experiments in the mixture experiments happened to yield a SLP; if both arms of the mixture experiment were actually performed, it is unlikely that the resulting data would yield a SLP. However, I do not see this as invalidating Birnbaum's argument. To quote Mayo (2014):

> Birnbaumization may be "performed" in the sense that T_B can be defined for any SLP pair $\mathbf{x}^*$, $\mathbf{y}^*$. Refer back to the hypothetical universe of SLP pairs, each imagined to have been generated from a θ-irrelevant mixture (Section 2.5). When we observe $\mathbf{y}^*$ we pluck the $\mathbf{x}^*$ companion needed for the argument. In short, we can Birnbaumize an experimental result: Constructing statistic T_B with the derived experiment E_B is the "performance."

5.4.1.5 Impact of Model Misspecification

An important caveat for L is that it assumes the model is correct, whereas in practice, to paraphrase George Box, "all models are wrong." This underlies the Bross quote in Section 5.1. As Berger and Wolpert (1988) write:

> The most frequently expressed criticism of the LP is that it is supposedly very dependent on assuming a particular parametric model with a density for X; since models are almost never known exactly, it is felt that the LP is only rarely applicable.

Berger and Wolpert (1988) counter that this criticism applies broadly to any statistical analysis. If a variety of alternative models are entertained, then L implies that the likelihood function corresponding to each model is the correct way to compare them. Others have noted that a Bayesian approach to robust inference embeds a set of possible models within a larger model that includes an indicator for each included model. Then, L can be applied to the larger model.

Mayo (2014) makes it clear that checks of the model do not violate L:

> An essential part of the statements of the principles [S, C and L] is that the validity of the model is granted as adequately representing the experimental conditions at hand [Birnbaum (1962), page 280]. Thus, accounts that adhere to the [L] are not thereby prevented from analyzing features of the data, such as residuals, in checking the validity of the statistical model itself.

The use of frequentist calculations for the purpose of model choice is a feature of the "Calibrated Bayes" approach to inference (Box 1980; Rubin 1984), which is the topic of Chapter 6.

5.4.2 Objections to the Weak Conditionality Principle

The indicator of choice of experiment Z in E_B is an ancillary statistic, so C is valid if it is considered appropriate to condition on ancillary statistics. This question has been previously discussed in Chapter 3; see in particular Cox's measuring instrument example, Example 3.2, which Cox uses to argue in favor of C. Durbin (1970) suggests that C does not apply if conditioning variables are restricted to depend solely on minimal sufficient statistics, but the justification for this restriction is not clear to me. Berger and Wolpert (1988) find the rejection of C counterintuitive and include a number of other compelling examples in support of this position. People who do not agree with conditioning on ancillary statistics need to address the engineer's question in the following example, provided by Pratt in his discussion of Birnbaum (1962):

> An engineer draws a random sample of electron tubes and measures the plate voltages under certain conditions with a very accurate volt-meter, accurate enough so that measurement error is negligible compared with the variability of the tubes. A statistician examines the measurements, which look normally distributed and vary from 75 to 99 volts with a mean of 87 and a standard deviation of 4. He makes the ordinary normal analysis, giving a confidence interval for the true mean. Later he visits the engineer's laboratory, and notices that the volt-meter used reads only as far as 100, so the population appears to be "censored." This necessitates a new analysis, if the statistician is orthodox. However, the engineer says he has another meter, equally accurate and reading to 1000 volts, which he would have used if any voltage had been over 100. This is a relief to the orthodox statistician, because it means the population was effectively uncensored after all. But the next day the engineer telephones and says, "I just discovered my high-range volt-meter was not working the day I did the experiment you analyzed for me." The statistician ascertains that the engineer would not have held up the experiment until the meter was fixed, and informs him

that a new analysis will be required. The engineer is astounded. He says,
"But the experiment turned out just the same as if the high-range meter
had been working. I obtained the precise voltages of my sample anyway,
so I learned exactly what I would have learned if the high-range meter had
been available. Next you'll be asking about my oscilloscope.

Pratt adds:

If the sample has voltages under 100, it doesn't matter whether the upper
limit of the meter is 100, 1000, or 1 million. The sample provides the same
information in any case. And this is true whether the end-product of the
analysis is an evidential interpretation, a working conclusion, a decision,
or an action.

5.4.3 Objections to the Sufficiency Principle

In my class discussions of L, students tend to question the validity of C and
accept S, perhaps because the notion of a sufficient statistic is so imbedded in
statistical theory, and Birnbaum (1962) states S as obviously true. However,
like Berger and Wolpert (1988), and as implied by Good's quote in Section 5.1,
I think S is more vulnerable to criticism than C, which seems to me intuitively valid. Consider the following example:

Example 5.3 (Example 5.1 continued)

Consider again the binomial vs. negative binomial experiment of
Example 5.1. As discussed in that example, the pair (x, n) of number of
successes and number of trials is a sufficient statistic. However, the distribution of (x, n) depends on the indicator Z for which experiment is
performed. The sufficiency principle goes beyond the definition of sufficiency, in that implies that the different distribution of (x, n) in the two
experiments does not matter for the inference. But that denies the premise of frequentist statistics, which is that the distribution does matter for
the inference. A frequentist can avoid L by simply denying S, without
rejecting the idea of a sufficient statistic.

Another way for a frequentist to deny L is simply to state that S does not
apply universally, and in particular does not apply for Birnbaum's mixture,
as argued by Kalbfleisch (1975). This weakening of S also has the benefit of
resolving the question of whether C and S are in fact contradictory, as discussed in Section 5.4.1.

5.5 Conclusion

If S is weakened as suggested in the previous subsection, frequentists still
need to examine arguments that C alone implies L. Indeed, acceptance of
the idea of conditioning on ancillaries creates undeniable difficulties for

frequentist inference. For example, there are examples where there are two ancillary statistics, say A and B, but (A,B) is not jointly ancillary. For frequentist inference, which ancillary statistic do we condition on? What do we do about "approximate ancillaries," as in the (2 × 2) table problem discussed in Chapter 2?

A different line of attack leading to L is that frequentist approaches to inference lead to incoherency from a decision-theoretic viewpoint, as argued by Berger and Wolpert (1982, Section 3.7). Coherency is a useful goal, to be sure, but not easily achieved in practice, when true models are unknown, uncertain assumptions are invoked, and model-checking is often based on the same set of data as that used to provide inference under the adopted model.

I leave it to you to decide if Birnbaum's ingenious (1962) argument emerges battered but intact from the various attacks to which it has been subjected. The Calibrated Bayes perspective, which is the topic of the next chapter, allows for frequentist approaches to model selection and checking, but is Bayesian for inference under the finally adopted model. This approach is in my view a satisfying fusion of frequentist and Bayesian ideas, and it adheres to the spirit (if not the letter) of the likelihood principle discussed here.

5.6 Some Thought Questions on This Chapter

1. "The likelihood principle more problematic for frequentists than for Bayesians." Discuss.
2. Do the arguments in favor of conditioning on ancillary statistics lead you to the conclusion that C is reasonable? If not, why not?
3. What is the definition of a sufficient statistic, and how does S go beyond this definition? Is the argument that S (together with C) implies L circular?
4. Which is more compelling to you, the frequentist position that the data-dependent stopping rule in Example 5.2 should matter, or Pratt's example in Section 5.4.2 implying that the stopping rule is irrelevant?

6

A Bayesian/Frequentist Compromise: Calibrated Bayes

The papers:

Box, G. E. P. (1980). Sampling and Bayes' inference in scientific modelling and robustness. *J. Roy. Statist. Soc. Series A*, 143, 4, 383–430.
Rubin, D. B. (1984). Bayesianly justifiable and relevant frequency calculations for the
applied statistician. *Ann. Statist.*, 12, 4, 1151–1172.

Other readings:

Little, R. J. (2006). Calibrated Bayes: a Bayes/frequentist roadmap. *Amer. Statist.*, 60,
3, 213–223.

6.1 Introduction

When I was a student of statistics in London in the early 1970s, debates raged about competing philosophies of statistical inference. Features of the debates included Birnbaum's (1962) "proof" of the likelihood principle, the topic of Chapter 5; books emphasizing issues of comparative inference (Barnett 1973; Cox & Hinkley 1974; Hacking 1965); read papers at the Royal Statistical Society that focused on competing systems of inference, with associated lively discussions (e.g. Bernado 1979; Dawid, Stone & Zidek 1973; Wilkinson 1977); claims of "counter-examples" to frequentist inference (Robinson 1975); and debates on the "foundations of survey inference," focusing on model-based versus design-based inference (Basu 1971; Hansen, Madow & Tepping 1983; Smith 1976), the topic of Chapter 13.

Then as now, statisticians can be roughly divided into one of three camps:

a. frequentists (F), who base inference for an unknown parameter θ on hypothesis tests or confidence intervals (CIs) or regions, derived from the distribution of statistics in repeated sampling.

b. Bayesians (B), who base inference about θ on its posterior distribution, under some model for the data and prior distribution for unknown parameters. A smaller set of "likelihoodists" try to develop inferences based on the likelihood without specifying a prior distribution.

c. Pragmatists, who do not have an overarching philosophy and pick and choose what seems to work for the problem at hand.

Since the 1970s, B has become much more practically feasible because of developments in Bayesian computation, the topic of Chapter 11. But it seems to me that the frequentist and pragmatist approaches to inference predominate. One might argue that good statisticians can get sensible answers under B or F; indeed, maybe two philosophies are better than one, since they provide more tools for the statistician's toolkit. For example, sampling statisticians often use randomization inference for some problems and models for other problems. I confess I am discomforted by this "inferential schizophrenia." Since B and F can differ even on simple problems, at some point, decisions seem needed as to which is right.

In his American Statistical Association Presidential Address at the Joint Statistical Meetings, Brad Efron wrote:

> The physicists I talked with were really bothered by our 250-year-old Bayesian–frequentist argument. Basically, there's only one way of doing physics, but there seems to be at least two ways to do statistics, and they don't always give the same answers. ... (Efron 2005).

Efron proposed empirical Bayes as a synthesis of Bayes and frequentist ideas, a topic that is taken up in Chapter 7 of this book. Empirical Bayes models place distributions on some parameters, and thus have a Bayesian flavor, but models are largely analyzed from the frequentist viewpoint, expanding the frequentist toolkit beyond the "fixed effects models" that were more prevalent at that time.

Jim Berger (2000) argues against the pragmatic approach to inference:

> Note that I am not arguing for an eclectic attitude toward statistics here; indeed, I think the general refusal in our field to strive for a unified perspective has been the single biggest impediment to its advancement.

Like Efron, Berger sees the need for synthesis, adding that "any unification that will be achieved will almost certainly have frequentist components to it."

Efron (2005) points out that frequentist and Bayesian approaches to inference both have strengths and weaknesses. So, it seems sensible to look for a compromise that combines the strengths of both paradigms. I see Calibrated Bayes, the topic of this chapter, as an attractive synthesis. The key idea is that inference for a particular dataset is based on a Bayesian model, but frequentist methods are applied to formulate, check, and refine the model, with the goal that resulting model-based inferences have good frequentist properties. Early proponents of the idea include Peers (1965),

Welch (1965), and Dawid (1982); I was drawn to the idea by the seminal papers by Box (1980) and Rubin (1984).

What does the term "calibrated" in Calibrated Bayes mean? To quote from Rubin (1984):

> ... what does the stated Bayesian probability, e.g. 95%, mean objectively or empirically? The question of tying the 95% to real world events is addressed here via the concept of frequency calibration. A Bayesian is calibrated if his probability statements have their asserted coverage in repeated experience. For example, if $\{I_1, I_2, ...\}$ represents a series of 95% Bayes interval estimates for unknowns $\{\theta_1, \theta_2, ...\}$ from known data sets $\{X_1, X_2, ...\}$, then these statements are calibrated if 95% of them cover their unknowns and 5% do not. A subsequence of $\{I_1, I_2, ...\}$ is calibrated if 95% of those I_j in the subsequence cover their unknowns. For an interesting discussion of this idea, see Dawid (1982). Clearly, it is desirable for a Bayesian to be calibrated overall and for all subsequences defined by characteristics of the data sets. If the Bayesian's models are correct, he will be calibrated overall and in all such subsequences. That is, if his models are correct, $\Pr(\theta_j \in I_j \mid X_j) = 0.95$ for all j, and thus averaging over all data sets $\Pr(\theta_j \in I_j) = 0.95$, or averaging over all data sets with observed characteristic $Q = Q(X_j)$, $\Pr(\theta_j \in I_j \mid X_j$ satisfies $Q) = 0.95$.

Thus, Calibrated Bayes aims to have the best of both worlds: the inference has all the good characteristics of Bayes, but the inference also has good frequentist properties. How is this objective achieved? An important feature is to allow frequentist calculations when developing the model. Both Box and Rubin make similar points about this frequentist role. Box (1980) writes:

> I believe that ... sampling theory is needed for exploration and ultimate criticism of the entertained model in the light of the current data, while Bayes' theory is needed for estimation of parameters conditional on adequacy of the model.

Rubin (1984) writes:

> frequency calculations are useful for making Bayesian statements scientific, scientific in the sense of capable of being shown wrong by empirical test; here the technique is the calibration of Bayesian probabilities to the frequencies of actual events.

In the next section, I review strengths and weaknesses of the Bayesian and frequentist paradigms, suggesting that Calibrated Bayes is a reasonable way of combining them. The description is based on Little (2006). In Section 6.3, I contrast the somewhat differing approaches to frequentist model checks espoused by Box and Rubin, indicating why I prefer Rubin's approach. I conclude in Section 6.4 with thoughts on the challenges in implementing the Calibrated Bayes approach.

6.2 Strengths and Weaknesses of the Frequentist and Bayesian Paradigms

6.2.1 Strengths of Frequentist Inference

In this section, I discuss some strengths of the frequentist approach to inference.

6.2.1.1 Frequentist Methods Do Not Need the Choice of a Prior Distribution

The frequentist paradigm avoids the need for a prior distribution and makes a clear separation of the role of prior information in model formulation and the role of data in estimating parameters. These pieces are treated on a more equal footing in the Bayesian approach, in that the prior density and likelihood are multiplied to create the posterior distribution.

6.2.1.2 Frequentist Methods Are Flexible

The frequentist approach is flexible, in the sense that full modeling is not necessarily required, and inferences lack the formal structure of Bayes Theorem under a fully specified prior and likelihood.

6.2.1.3 Frequentists Seek Methods with Good Operating Characteristics

The focus on repeated sampling properties tends to ensure that frequentist inferences are well calibrated; for example, in the survey sampling setting, design-based inference automatically takes into account survey design features that might be ignored in a model-based approach.

6.2.2 Weaknesses of Frequentist Inference

6.2.2.1 Frequentist Theory Is Not Prescriptive

There is no unified theory for how to generate frequentist inferences; frequentist theory is more a set of concepts for assessing properties of inference procedures rather than an inferential system per se.

More specifically, the principle of least squares is too limited for a general theory. Unbiasedness seems a desirable property, but it has severe limitations as a general principle, as discussed in Chapter 7. Generalized estimating equations, discussed in Chapter 10, provide a very broad class of procedures, but there seems no general prescription for how to choose the equations, and the theory is basically asymptotic.

While frequentist inference as a whole is not prescriptive, some parts of it are: in particular, Efron (1986) suggests that Fisher's theory of maximum likelihood (ML) estimation, with measures of uncertainty based on the observed information, is popular since it provides an "automatic" form of frequentist inference. Efron (1986) contrasts this with Bayesian theory, which "requires a great deal of thought about the given situation to apply sensibly." However, in many situations, Bayes with a reference prior (RP) distribution often improves on ML inferences with small samples and is similar to ML inference with large samples. In his discussion of Efron (1986), Lindley notes the automatic nature of Bayes Theorem for generating the posterior distribution.

6.2.2.2 Exact Small-sample Tests and Confidence Intervals Are not Available for Many Statistical Models

Exact frequentist properties – for example, a 95% CI covering the true parameter 95% of the time in repeated sampling – are not available for many statistical models. The Behrens-Fisher problem discussed in Chapter 5 is one famous example. Many theoretical properties are asymptotic and are of limited use for assessing the trade-off between model complexity and the available sample size.

6.2.2.3 Frequentist Methods Are Ambiguous about Conditioning, and Violate the Likelihood Principle

Specifically, the reference set for determining repeated-sampling properties is often ambiguous, and frequentist theory suffers from non-uniqueness of ancillary statistics (Cox 1971), as discussed in Chapters 3 and 5.

6.2.2.4 Good Operating Characteristics in Repeated Sampling Do Not Necessary Imply that an Inference Is Appropriate for the Data Set Actually Observed

Good frequentist properties are arguably necessary to make an inference "scientifically justified," but they are not sufficient, if the analysis does not condition on all the relevant information. Bayesian inferences do this by construction, because they condition on all the data. Rubin (1984) uses the term "Bayesianly justifiable" for this property. Little (2006) gives the following example:

Example 6.1: Single sample t inference with a bound on precision.

Consider an independent normally distributed sample with $n = 7$ observations, with sample mean $\bar{x} = 1$ and standard deviation $s = 1$. If the population standard deviation (say σ) is unknown, the usual t-based 95% interval for the population mean is

$$I_{.05}^{BRP}(s) = I_{.05}^{F}(s) = \bar{x} \pm 2.447 \left(s/\sqrt{n} \right) = 1 \pm 0.92, \qquad (6.1)$$

where a frequentist F interprets this as a 95% CI, and a Bayesian B interprets it as a 95% posterior credibility interval, based on Jeffreys' reference prior (RP) distribution $p(\mu,\sigma) \propto 1/\sigma$. The correspondence of the B and F intervals is well known.

Suppose now that we are told that $\sigma = 1.5$, as when σ is the known precision of a measuring instrument. The standard 95% interval then replaces s by σ and the t by the normal percentile:

$$I_{.05}^{BRP}(\sigma = 1.5) = I_{.05}^{F}(\sigma = 1.5) = \bar{x} \pm 1.96\left(1.5/\sqrt{n}\right) = 1 \pm 1.11. \qquad (6.2)$$

Both (6.1) and (6.2) have exact 95% confidence coverage. A collaborator hoping for an interval that excludes a null value of zero might prefer (6.1), but we can all agree that (6.2) is the more appropriate inference, the wider interval reflecting the information that the sample variance s is underestimating the true variance σ.

Now, suppose the experimenter remembers some additional unaccounted sources of variability, implying that $\sigma > 1.5$. Three candidate 95% intervals for μ are Eqs. (6.1) and (6.2), or the Bayesian credibility interval with the RP modified to incorporate the constraint that $\sigma > 1.5$, namely:

$$I_{.05}^{BRP}(\sigma > 1.5) = 1 \pm 1.45. \qquad (6.3)$$

Pick your poison:

a. Eq. (6.1) seems the optimal 95% frequentist CI, given that it is has exact nominal coverage and σ is unknown, but it is not appropriate for the data set at hand: how can the knowledge that $\sigma > 1.5$ leads to a narrower interval than Eq. (6.2), the standard frequentist interval when $\sigma = 1.5$? The problem is that the interval ignores the lower bound on σ, but only Bayesian methods seem capable of appropriate accounting for this information.

b. Eq. (6.2) is the obvious asymptotic approximation, given that 1.5 is the ML estimate of σ. However, the appeal to asymptotics is clearly inappropriate for $n = 7$, and it is not clear with what to replace it. The constraint $\sigma > 1.5$ implies that Eq. (6.2) has less than 95% confidence coverage, since it is based on a known underestimate of σ. Interestingly, this is true even though it contains the exact t CI, Eq. (6.1), for the observed sample. This could not happen for a Bayesian posterior credible interval, for which a wider interval must have higher posterior probability.

c. Eq. (6.3) is the Bayes' interval subject to the constraint that $\sigma > 1.5$. It appropriately wider than Eq. (6.2), perhaps too wide, reflecting the choice of prior, but it is (as always) dependent on the choice of prior distribution.

6.2.3 Strengths of the Bayesian Approach

6.2.3.1 Users Interpret Answers Bayesianly, and Bayes Gives the Right Inference if the Model Is Well-Specified

Rubin (1984) writes:

> ...consumers of statistical answers, at least interval estimates, almost uniformly interpret them Bayesianly, that is as probability statements about the likely values of parameters. Consequently, the answers statisticians provide to consumers should be capable of being interpreted as approximate Bayesian statements - that is, statisticians' summary statements should be Bayesianly justifiable in the sense defined in Section 1.

Bayes gives a coherent inference, in the sense that Bayes Theorem is the correct way to update beliefs (as represented by probability distributions) to incorporate new information (e.g. de Finetti 1974; Savage 1954). See also Berger and Wolpert (1988) and the discussion in Chapter 5.

6.2.3.2 Bayesian Models, by Treating Parameters as Random, Tend to Lead to Better Answers

Rubin (1984) shares Efron's (2005) enthusiasm for mixed-effects models, writing:

> Since Bayesian models treat all unknowns as random variables, Bayesian models formulate distributions for parameters, and thereby naturally create models with multiple levels of randomness. Commonly, the answers that are derived from such models with levels of randomness are termed empirical Bayesian, but the essential feature here is not the label attached to the estimators but the Bayesian structure, which treats the unknown parameters of interest as random variables. The resultant extra flexibility generally leads to better answers by allowing borrowing of strength.

An example is small-area estimation from survey data. Unlike empirical Bayes, fully Bayesian models require specification of a prior distribution for variance components, but the result is answers that better propagate uncertainty in estimation. This point is discussed further in Chapter 7.

6.2.3.3 Bayes Gives Answers to Complex Problems Where Frequentist Inferences Are Asymptotic

The development of Monte Carlo computational methods (discussed in Chapter 11) means that Bayes is now practically feasible for complex models involving large numbers of parameters. Unlike frequentist inferences which are asymptotic except for a narrow range of problems, Bayesian answers can provide small-sample inferences that take into account uncertainty in parameters; in particular, integrating over nuisance parameters is in my view the best way to propagate uncertainty arising from the fact that they are not known.

6.2.4 Weaknesses of Bayesian Inference

Some criticisms of Bayesian methods are better justified than others. I present some of them here, concluding with the most important one.

6.2.4.1 *Bayes Requires and Relies on Full Specification of a Model (Likelihood and Prior)*

In general, Bayes involves a full probability model for the data and includes a prior distribution for unknown parameters – Efron (2005) comments on this high degree of specification. Developing a good model is often challenging, particularly in complex problems. Where does this model come from? Is it trustworthy? Bayes' is much less prescriptive about how to select models than it is once model and prior distribution are selected.

6.2.4.2 *Bayesian Inference Is too Subjective*

A common criticism is that Bayes is too subjective for a scientific inference, requiring a subjective definition of probability and the selection of a prior distribution. However, subjectivity is a matter of degree, and Bayesian models can run the full gamut, from standard regression models with RPs that yield posterior credible inferences that mimic frequentist CIs, to more subjective models that bring in substantial prior information through a proper prior distribution. A broad view of Bayesian methods includes methods based on weakly informative priors that some classify as frequentist (e.g., Samaniego & Reneau 1994). Frequentist methods also vary in subjectivity. For example, the coefficient of a covariate omitted from a regression equation is in effect being given a sharp prior distribution with all the probability at zero. Models with strong assumptions, such as models for selection bias (e.g., Heckman 1976), are no less subjective because they are analyzed using frequentist methods. Some statisticians worry about the subjective definition of probability that underlies the Bayesian approach, but I am not one of them.

6.2.4.3 *Bayes Discounts the Importance of Randomization*

Another criticism of Bayesianism that I find unjustified is that it denies the role of randomization for design, since the randomization distribution is not the basis for model-based inferences. Indeed, some Bayesians have stoked this criticism by denying that randomization plays any kind of useful design role. On the contrary, the utility of randomization from the Bayesian perspective becomes clear when the model is expanded to include indicators for the selection of cases or allocation of treatment (Gelman et al 2003; Rubin 1978,

Sections 3 and 4). Randomization provides a practical way to assure that the selection or allocation mechanisms are ignorable for inference, without making ignorable selection or allocation a questionable assumption. I discuss this point in more detail in Chapters 13 and 14.

6.2.4.4 Bayes Yields "Too Many Answers"

If the frequentist paradigm does not provide enough exact answers, with Bayes, there is an embarrassment of riches. Once the likelihood is nailed down, every prior distribution leads to a different answer! If forced to pick a prior distribution, the problem is which prior to choose. If the mapping from the prior distribution to the posterior distribution is considered the key, as argued cogently by some Bayesians (e.g., Leamer 1978), there is still a problem with the surfeit of posterior distributions. Sensitivity analysis is often a rational choice, but it is not a choice that appeals much to practitioners who are looking for clear-cut answers.

6.2.4.5 Models Are Always Wrong, and Bad Models Lead to Bad Answers

The main problem with Bayes is that models are at best simplified idealizations, and models that are seriously wrong lead to poor answers. This makes the search for good model checks important. Two Bayesian approaches to this are (a) to embed the assumed model within a richer model that relaxes one of more of its assumptions, or (b) to compare alternative models using a Bayesian hypothesis test. Bayesian model averaging is intuitive and compelling, but in any given problem, there is still the problem of deciding the class of models over which averaging takes place, and how to choose the prior probabilities of models in the class. Rubin (1984) writes:

> The use of convenient statistics, such as residuals, to monitor models is now an established part of sound statistical practice. An alternative to examining such statistics is to embed the current model in a richer and richer web of Bayesian models. Although possible in principle, this seems to be a hopelessly complex task to always implement in practice. It is usually substantially more difficult to create a new relevant model and perform a full Bayesian under it than to check the distribution of a few cleverly chosen statistics. Certainly, when there is no indication that the current model is inadequate, the extra modeling effort may often be a poor expenditure of time.

Bayesian hypothesis testing has the logic of Bayes Theorem working in its favor, but comparing models of different dimension is tricky and sensitive to the choice of priors. To quote from Gelman and Shalizi (2013):

> The main point where we disagree with many Bayesians is that we do not see Bayesian methods as generally useful for giving the posterior probability that a model is true, or the probability for preferring model A over model B, or whatever. Beyond the philosophical difficulties, there

are technical problems with methods that purport to determine the posterior probability of models, most notably that in models with continuous parameters, aspects of the model that have essentially no effect on posterior inferences *within* a model can have huge effects on the comparison of posterior probability *among* models. Bayesian inference is good for deductive inference within a model, we prefer to evaluate a model by comparing it to data.

Given these considerations, it seems that Bayes yields useful model assessment tools but is unlikely to achieve the degree of clarity of Bayesian inference under an agreed model.

6.3 Calibrated Bayes: Combining the Strengths of Bayes and Frequentist Approaches

6.3.1 Overview

A crude assessment of the strengths and weaknesses described above is that Bayes is relatively strong for inference under an assumed model but less strong for model formulation and assessment, and frequentist methods are potentially weak for inference on a particular data set but relatively strong for model assessment. If this summary is accepted, then the natural compromise is to use frequentist methods for model development and assessment, and Bayesian methods for inference under a model. This is the Calibrated Bayes perspective. Fisherian significance tests have a role within the Calibrated Bayes paradigm, for model checking (Box 1980; Rubin 1984). For example, a global test of whether data are consistent with a null model is allowed without the need to specify the alternative hypothesis. On the other hand, classical hypothesis testing does not have a role for inference about model parameters – not in my view a serious loss.

6.3.2 Prior and Posterior Predictive Checks

Standard frequentist model assessment is based on statistics that are independent of the model parameters, such as plots of standardized residuals in regression. Box (1980) and Rubin (1984) extend these methods to allow checking statistics that depend on the parameters, but their methods differ in important respects. Box's methods are called prior predictive checks and Rubin's methods are called posterior predictive checks. Both approaches involve choosing a scalar discrepancy function $d(Y)$ of the data Y, and then assessing how extreme is the observed value of $d(Y)$, say $d(Y_{obs})$, with respect to the distribution of $d(Y)$ over repeated sampling of the data. The use of repeated sampling for checking the model does not violate the likelihood principle, which assumes the truth of the hypothesized model. It is also consistent with the philosophy of assessing the operating characteristics of the assumed model.

Box (1980) bases his implementation of this idea on the factorization:

$$p(Y, \theta \mid M) = p(Y \mid M)p(\theta \mid Y, M), \tag{6.4}$$

where the second term on the right side is the posterior distribution of the parameter θ given data Y and model M, and is the basis for inference, and the first term on the right side is the marginal distribution of the data Y under the model M, and is used to assess the validity of M.

Specifically, discrepancy functions of the observed data $d(Y_{obs})$ are assessed from the perspective of realizations from their marginal distribution $p(d(Y) \mid M)$. The factorization (6.4) has intuitive appeal, but a questionable feature of this "prior predictive checking" is that checks are sensitive to the choice of prior distribution, even when this choice has a limited impact on the posterior inference; in particular, it does not provide checks functions with distributions that depend on parameters with weakly informative priors where $p(d(Y) \mid M)$ is not a well-defined distribution.

Posterior predictive checks (Gelman, Meng & Stern 1996; Rubin 1984) do not suffer from this drawback. They are based on a different factorization than that of Box (1980), namely:

$$p(Y^*, \theta^*, \theta \mid Y_{obs}, M) = p(Y^*, \theta^* \mid Y_{obs}, \theta, M)p(\theta \mid Y_{obs}, M), \tag{6.5}$$

where (Y^*, θ^*) is the realization of a future data and parameter values based on the posterior predictive distribution given model M and observed data Y_{obs}. This leads to posterior predictive checks, which compare the checking function with values of its posterior distribution given M and Y_{obs}. These checks involve an amalgam of Bayesian and frequentist ideas but are clearly frequentist in spirit, because they embed the observed data within a sequence of unobserved data sets that could have been generated under M, and check whether the observed data are "extreme."

Posterior predictive checks have been criticized for using the data twice and not yielding tail-area P-values that are uniformly distributed under the posited model (Bayarri & Berger 2000; Robins, van der Vaart & Ventura 2000). The data generated under the model M and observed data Y_{obs} are "more dispersed" than the observed data because they do not fix the value of θ, but rather generate data sets that integrate over the posterior distribution of θ. As a result, posterior predictive checks are conservative, particularly if the posterior distribution of θ is dispersed. That is, a low P-value is indicative of model misspecification, but a P-value that is not low does not necessarily suggest that the model is valid. Whether this deviation from frequentist properties is important is a matter of some controversy and debate. For example, Gelman (2013) presents two examples that yield opposite conclusions about this question.

Generating data under the model with θ fixed at some value, such as the ML estimate $\hat{\theta}$, yields data more similar to the actual data but does not allow for uncertainty in estimating the true value of θ under the hypothesized model. Bayarri and Berger (2000) describe limitations of this method, calling the resulting statistics "plug-in P-values." They propose two modifications of posterior predictive P-values, namely *partial posterior predictive P-values* and *conditional predictive P-values*, which modify the posterior distribution of θ in posterior predictive checks by conditioning on statistics to alleviate the problem of "double dipping."

In particular, consider data x for a model $f(x|\theta)$ with prior distribution $\pi(\theta)$, observed data x_{obs}, and a checking statistic T with observed data value t_{obs}. The partial posterior predictive P-value has the form

$$p_{\text{ppost}} = \text{Pr}^{m(\cdot|x_{\text{obs}}\setminus t_{\text{obs}})}(T \geq t_{\text{obs}}),$$

where

$$m(t \mid x_{\text{obs}}\setminus t_{\text{obs}}) = \int f(t\mid\theta)\pi(\theta\mid x_{\text{obs}}\setminus t_{\text{obs}})$$

and $\pi(\theta\mid x_{\text{obs}}\setminus t_{\text{obs}})$ is the (assumed proper) partial posterior distribution of θ:

$$\pi(\theta\mid x_{\text{obs}}\setminus t_{\text{obs}}) \propto f(x_{\text{obs}}\mid t_{\text{obs}},\theta)\pi(\theta) \propto \frac{f(x_{\text{obs}}\mid\theta)\pi(\theta)}{f(t_{\text{obs}}\mid\theta)}.$$

Bayarri and Berger (2000) provide examples suggesting that the resulting checks have a distribution closer to uniform under the posited model. Robins, van der Vaart and Ventura (2000) also propose modified methods that yield asymptotically uniform distributions of P-values under the posited model.

6.4 Examples

Example 6.2. The Behrens-Fisher problem

In Chapter 5, we considered inferences for the Behrens-Fisher problem, where the data consist of two independent normal samples:

$$(x_{i1}, i = 1,...,n_1) \sim_{\text{iid}} G(\mu_1,\sigma_1^2), (x_{i2}, i = 1,...,n_2) \sim_{\text{iid}} G(\mu_2,\sigma_2^2); \theta = (\mu_1,\sigma_1^2,\mu_2,\sigma_2^2),$$

where $G(\mu, \sigma^2)$ denotes the normal (Gaussian) distribution with mean μ and variance σ^2. The goal is an interval estimate for the differences in means $\delta = \mu_2 - \mu_1$. Two possible solutions are:

a. the $100(1-\alpha)\%$ posterior credible interval for δ, assuming a Jeffreys' prior on the parameters, namely

$$P(\mu_1, \mu_2, \sigma_1, \sigma_2) \propto \sigma^{-1}\sigma_2^{-1}, \qquad (6.6)$$

which is also the fiducial solution favored by Fisher (1935).

b. the $100(1-\alpha)\%$ posterior credible interval for δ, assuming the prior distribution:

$$p(\mu_1, \mu_2, \sigma_1, \sigma_2) = \sigma_1^{-3}\sigma_2^{-3}(\sigma_1^2/n_1 + \sigma_2^2/n_2), \qquad (6.7)$$

the solution proposed by Ghosh and Kim (2001). As discussed in Section 4.3, the Ghosh & Kim solution is better calibrated in repeated sampling, because it is second-order probability matching, whereas the credible interval from (6.6) is only first-order probability matching. Thus, from the Calibrated Bayes perspective of this chapter, the Ghosh & Kim interval is preferred; this despite the fact that from a purely subjective Bayes perspective, the prior distribution (6.7) is questionable because it depends on the sample sizes.

Example 6.3. Sequential analysis: stopping rules leading to less robust inference

Clinical trials often have interim analyses to assess whether they should be continued or stopped because one treatment has a decisive advantage. Should inferences be affected by these multiple looks at the data? A frequentist says yes, tests need to be modified to maintain the nominal alpha-level for tests of the null (spending functions) (e.g., DeMets & Ware 1980.) A Bayesian says no, the stopping rule is ignorable, so the posterior distribution is unaffected by prior looks at data (e.g., Lindley 1972, p. 24). This example is cited as a counterexample by both Bayesians and frequentists! If we statisticians can't agree which theory this example is counter to, what is a clinician to make of this debate?

Since the Calibrated Bayes inference is Bayesian, there are no penalties for peeking in the inference – the inference is unaffected by interim analysis. On the other hand, interim analyses and stopping rules do increase sensitivity of the Bayesian inference to the choice of prior distribution (Rosenbaum & Rubin 1984), so they do have subtle implications for the robustness of the inference. Thus, models for data sets subject to interim analyses need to be carefully justified and checked.

Example 6.4. Tests of independence in a 2×2 contingency table

The standard Bayesian inference adds Beta priors for the success rates in the two groups and computes the posterior probability that $\pi_1 > \pi_2$.

Proper prior distributions may be entertained in certain contexts; when there is little prior evidence about the success rates, the choice of "objective prior" has been debated, but Jeffreys' prior is one plausible conventional choice. The Fisher exact test P-value corresponds to an odd choice of prior distribution (Altham 1969). As discussed in Chapter 2. Howard (1998) argues for alternative choices. Calibrated Bayes methods limit ambiguities in the reference set for frequentist assessments to model evaluation, rather than to model inference under a specified model. In particular, Gelman (2003) argues for posterior predictive checks that condition on the margin of the contingency table fixed by the design.

Example 6.5. Survey samples with complex designs

In the sample survey context discussed in Chapter 13, the Calibrated Bayes perspective suggests models that include key features of the sample design, such as stratification, sample weights, and clustering induced by multistage sampling (Little 2012, 2022.) Critiques of the model-based approach (e.g., Hansen, Madow & Tepping 1983) consider models that do not do this, which explains their poor properties in repeated sampling.

6.5 Conclusions

Bayesian models have the flexibility to handle simple and complex situations. Because the posterior distribution conditions on the data, issues of how to deal with ancillary statistics do not arise, and results are what Rubin (1984) calls "Bayesianly justified." The other side of the Calibrated Bayes coin is that models are imperfect idealizations, and hence need careful checking; this is where frequentist methods have an important role. These methods include Fisherian significance tests of null models, diagnostics that check the model in directions that are important for the target inferences, and model-checking devices like posterior predictive checking and cross-validation.

Calibrated Bayes does not solve all the problems of statistical inference. Ambiguities arise at the frontier between model inference and model checking. How much peeking at the data is allowed in developing the model without seriously corrupting the inference? When is model selection appropriate as opposed to model averaging (e.g. Draper 1995)? There remains much to argue about here, but I still think that the Calibrated Bayes provides a useful roadmap for many problems of statistical modeling and inference.

6.6 Some Thought Questions on This Chapter

1. For a given model, wider credible intervals have higher posterior probability. This property does not necessarily hold for CIs. Describe why Example 6.1 provides an example of this.

2. To what extent does Calibrated Bayes align with the Fisherian and Neyman/Pearson approaches to hypothesis testing?

3. "Calibrated Bayes is consistent with the likelihood principle, even though it entertains frequentist model checks." Explain.

4. "From the Calibrated Bayes perspective, Bayesians should be frequentist, and frequentists should be Bayesian." Discuss how Calibrated Bayes attempts to realize this idea.

5. What are the relative merits of prior and posterior probability checks for assessing the model?

6. Posterior predictive P-values are not uniformly distributed even if the model is correct. Review this critique of posterior predictive P-values and the extent to which this invalidates this form of model check.

Part II

Statistical Methods

7

Baseball Averages, Foreign Cars, and Shrinkage Estimation

The paper:

Efron, B. & Morris, C. (1977). Stein's paradox in statistics. *Scientific American*, 236, 5, 119–127.

Other readings:

Efron, B. & Morris, C. (1973). Stein's estimation rule and its competitors—an empirical Bayes approach. *J. Amer. Statist. Assoc.*, 68, 341, 117–130.

Rubin, D. B. (1980). Using empirical Bayes techniques in the law school validity studies (with discussion). *J. Amer. Statist. Assoc.*, 75, 372, 801–881.

Shen, W. & Louis, T. A. (1998). Triple-goal estimates in two-stage hierarchical models. *J Roy. Statist. Soc. Ser B*, 60, 2, 455–471.

7.1 Introduction

Efron and Morris (1977) concern a seminal idea in statistics, empirical Bayes shrinkage. As the son of a tabloid journalist who loved simple non-verbose language, I admire the clear and enjoyable description of statistical ideas for a nonstatistical audience.

As other readings, I include Efron and Morris (1973) for more technical details, and the Rubin (1980) paper because it is an interesting application of shrinkage to the transformed ratios of two regression coefficients, and the discussion raises interesting issues. Shen and Louis (1998) give a useful account of the Bayesian approach to shrinkage estimation via hierarchical models, and in particular the role of the loss function in determining summaries of the posterior distribution.

The central idea of Efron and Morris (1977) is "shrinkage," specifically, estimators obtained by shrinkage of the deviations of the individual means from the overall mean. The sample mean is the best unbiased estimate of the mean for a random sample from a normal population, in terms of mean squared error loss. It is also intuitively sensible – why would one not use a sample mean to estimate a population mean? But James and Stein (1961) showed that estimating three or more normal means by their sample means is suboptimal, and a better procedure "shrinks" the individual sample means toward

DOI: 10.1201/9781003395164-9

zero. The resulting estimates are no longer unbiased for their respective population means but are superior as predictors. This attack on unbiased estimation caused a considerable stir, and as Efron and Morris indicate, created a considerable backlash.

Efron and Morris (1977) refine the James-Stein estimator by shrinking toward the overall mean rather than towards zero. As a result of papers like Efron and Morris (1977), unbiasedness lost some of its shine as a statistical property. The paper suggests two distinct routes to shrinkage estimation, reflecting somewhat different statistical philosophies:

a. James and Stein (1961) arrived at their estimator through decision theory, showing that under certain conditions, in problems involving at least three normal means, sample mean estimates are inadmissible with respect to square error loss; that is, they are uniformly inferior to the James-Stein shrinkage estimator for all values of the underlying parameters.

b. The other route to shrinkage is via random-effects or mixed-effects models, which posits a statistical model for the data that treats at least some parameters as random rather than fixed quantities.

Concerning (a), mathematics plays a central role in statistical theory, and decision theory is a very important part of statistics. However, I favor the statistical modeling approach to shrinkage, because I think modeling is at the heart of good statistical practice. I like models, and their associated assumptions, because they make clear the conditions under which statistical inferences can be justified or, if suspect, might be improved. To me, many "assumption-free" methods are in fact "assumption-hidden" methods. We can weaken parametric assumptions, but only up to a point.

A more concrete reason for preferring the modeling approach is the so-called "Stein's Paradox" that appears in the title of Efron and Morris (1977) and is described below. I think the so-called "paradox" is resolved by taking a modeling perspective, a point in favor of the modeling approach.

I set the stage for discussing the ideas in Efron and Morris (1977) with three examples, the first two of which play an important role in their paper, and the third example my own.

Example 7.1. The baseball average example

Efron and Morris (1977) consider the batting averages – number of hits divided by number of times at-bat – of $k = 18$ Major-League players after their first 45 times at-bat in the 1970 season. These averages are the gray bars in Figure 7.1 and range from 0.152 to 0.400. The middle bars show the averages at the end of the season, which are based on a large enough number of bats to approximate the "true" averages. Clearly, these

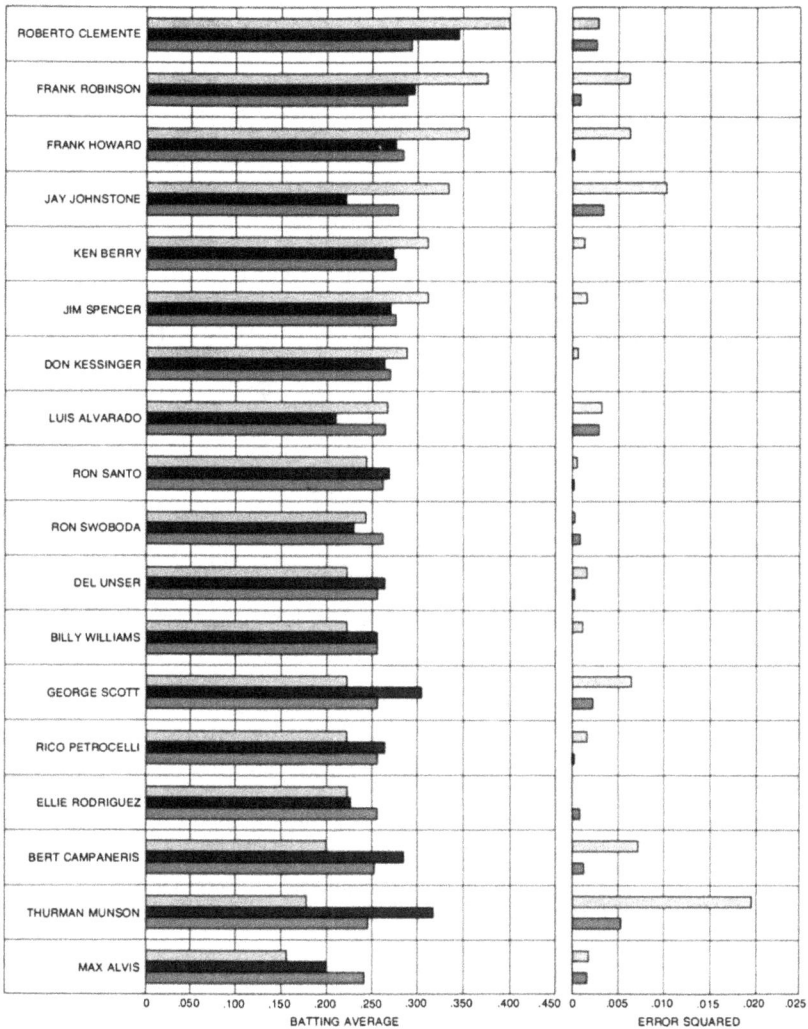

FIGURE 7.1
Batting averages of 18 baseball players. The upper bars are sample means after 45 at-bats, the middle bars are the averages at the end of the season, and the lower bars are the James-Stein estimates. The right panel indicates that the James-Stein estimates are closer to the averages at the end of the season for 16 of the 18 players. (Reproduced with the permission of Dr. Efron from Efron and Morris (1977), p. 120.)

averages are less dispersed – the lowest one (0.152) is increased to 0.200, and the highest one (0.400) reduced to 0.349. This illustrates the "regression to the mean" effect. The James-Stein estimator of the true averages for player i is

$$\bar{y}_{JS,i} = \bar{y} + c(\bar{y}_i - \bar{y}), \tag{7.1}$$

where $\bar{y}_i$ is the sample average after 45 at-bats, $\bar{y} = \sum_{i=1}^{k} \bar{y}_i / N$ is the grand mean, namely 0.265, and c is the shrinkage factor, computed as

$$c = 1 - \frac{(k-3)\sigma^2}{\sum_{i=1}^{k}(\bar{y}_i - \bar{y})^2},$$ (7.2)

where $\sigma^2 = \bar{y}(1 - \bar{y})/45$ is an estimate of the variance of the sample means. The lower of the three bars for each player displays these estimates.

The "error-squared" bars on the right side of Figure 7.1 show the squared deviations of the initial averages and James-Stein estimates from the end-of-season averages. The James-Stein estimates are closer to the "true" averages, that is, yield better predictions, for 16 of the 18 players. As Efron and Morris describe it, the James-Stein estimates "anticipate" the regression-to-the-mean effect.

Example 7.2. Cars in Chicago are added to the baseball averages in Example 7.1

Suppose that an apparently unrelated data point is added to the data set in Example 7.1, namely the proportion of a random sample of 45 cars in Chicago that are imported. The decision-theoretic contention that the James-Stein estimate is superior still applies to the expanded data set, because it makes no particular assumption about the means in question. This suggests that the James-Stein estimates still improve on the sample means for the augmented data set. This is the "Stein Paradox" in 'Efron and Morris's (1977) title: why should the cars be included, given that the percentage of foreign cars and baseball averages bear no relationship whatsoever with each other, and intuitively should not be combined.

Example 7.3. The application of shrinkage to a set of diverse statistical consulting problems

A consulting statistician has data from 18 unrelated consulting projects carried out in the course of a year that all involve estimating a population mean, with sample sizes ranging from 45 to 450. A colleague, on reading Efron and Morris (1977), advises replacing the sample means, which the consulting statistician has used to estimate the target means by the James-Stein estimates, arguing that according to decision theory, this will improve the estimates. Here, the sample sizes are not the same across the projects, so the application is closer to the second example in Efron and Morris (1977) concerning disease toxoplasmosis in El Salvador. But, as Efron & Morris describe, the James-Stein estimator can be readily modified for that situation; means with greater standard errors receive higher shrinkage factors than means with smaller standard errors, as is appropriate given their greater uncertainty.

Should the consulting statistician apply shrinkage in this setting? Decision theory suggests that the sample means are inadmissible, but why should estimates be influenced by results from completely unrelated problems?

7.2 Alternative Rationales of Shrinkage Estimation

7.2.1 The Decision-Theoretic Argument

I consider the more general situation where the mean $\bar{y}_i$ is based on a sample size n_i that is not necessarily constant across i. Underlying 'Efron and Morris's (1977) decision-theoretic argument is the normal model:

$$\bar{y}_i \sim_{\text{ind}} N(\mu_i, \delta^2/n_i), \tag{7.3}$$

(see Eq. (1.1) in Efron & Morris 1972) wherein Example 7.1, $n_i = 45$ and the variance $\sigma^2 = \delta^2/45$ is the same for all i. No assumptions are made about the means $\{\mu_i\}$, which are fixed parameters. The argument assumes an exchangeable quadratic loss function

$$L(\hat{\mu}_1, ..., \hat{\mu}_k \mid \mu_1, ..., \mu_k) = \sum_{i-1}^{k} (\hat{\mu}_i - \mu_i)^2,$$

where $\hat{\mu}_i$ is the estimate of μ_i. The risk function is then the expected loss:

$$R(\hat{\mu}_1, ..., \hat{\mu}_k \mid \mu_1, ..., \mu_k) = \sum_{i-1}^{k} E\left((\hat{\mu}_i - \mu_i)^2 \mid \mu_1, ..., \mu_k\right). \tag{7.4}$$

For problems with single mean, $k = 1$, the sample mean is admissible, in the sense that no alternative estimate has a lower risk for all values of the parameters. But when $k > 2$ the sample means $\hat{\mu}_i = \bar{y}_i, i = 1, ...k$ are inadmissible, because under exchangeable quadratic loss, the James-Stein estimates have uniformly lower risk for all values of the parameters. James-Stein shrinkage may not work for other loss functions; for example, shrinkage to zero can perform poorly if the quantity of interest is the maximum expected squared loss over all the means.

As applied to the baseball example, the model Eq. (7.3) has deficiencies. The baseball averages $\bar{y}_i$ are proportions, so the binomial distribution would be more appropriate than the normal distribution. Efron and Morris (1977) justify normality by appealing to the Central Limit Theorem, which is perhaps an acceptable approximation given the sample size of 45 at-bats. Also, it would be preferable to replace the variance δ_i^2/n_i in (7.3) by the binomial variance expression $\mu_i(1-\mu_i)/n_i$, which depends on the underlying true mean μ_i, and is not a constant σ^2 for all i when n_i is constant. For the range of true means in the baseball example, this is not a terrible approximation, and the evidence of Figure 7.1 suggests that the underlying method still works well despite these deficiencies. An improvement would be to apply an arc-sine transformation to the proportions.

Perhaps a more intriguing feature of (7.3) is that it describes a "fixed-effects" model, in the terminology of analysis of variance (ANOVA); the means μ_i are treated as fixed parameters. The "Stein paradox" arises because, as Efron and Morris (1977) state (my italics):

> Stein's theorem is concerned with the estimation of several unknown means. *No relation between the means* need be assumed: they can be batting abilities or proportions of imported cars.

This argument appears to justify including the imported cars with the batting averages when computing James-Stein estimates in Example 7.2. But from a modeling perspective, the fixed-effects model (7.3) does *not* lead to shrinkage of the means – the maximum likelihood (ML) estimate of μ_i is the sample mean $\bar{y}_i$, which has the optimal asymptotic properties of ML discussed in Chapter 1, notwithstanding the admissibility argument. Shrinkage is justified by the random-effects model that I now discuss.

7.2.2 The Random-Effects Model

Consider the one-way random-effects ANOVA model:

$$(y_{ij} \mid \mu_i, \delta^2, \tau^2) \sim N(\mu_i, \delta^2), i = 1,\dots,k, j = 1,\dots,n_i$$
$$(\mu_i \mid \delta^2, \tau^2) \sim N(\mu, \tau^2),$$
(7.5)

the normal model (7.3) with the additional assumption that the means $\{\mu_i\}$ are sampled from an underlying normal distribution. Given the parameters (μ, δ^2, τ^2), the best estimate of μ_i is its conditional mean given $\bar{y}_i$, which is found by an application of Bayes Theorem to be

$$E(\mu_i \mid \bar{y}_i, \mu, \tau^2, \delta^2) = \mu + c'(\bar{y}_i - \mu), \text{ where } c' = \frac{\tau^2}{\tau^2 + \delta^2/n_i}.$$
(7.6)

This has a similar shrinkage form to Eq. (7.1), while also allowing for unequal sample sizes $\{n_i\}$. If τ^2 is large compared with δ^2, then c' is close to one and the degree of shrinkage is minimal.

In practice, the parameters (μ, δ^2, τ^2) are unknown. In empirical Bayes estimation, they are estimated, either by ML (which involves an iterative algorithm) or by the method of moments (which is noniterative but less efficient; see, e.g., Carter & Rolph 1974). Alternatively, the parameters can be assigned a prior distribution, and μ_i estimated by its posterior mean $E(\mu_i \mid \bar{y}_i)$, an approach that propagates uncertainty in the parameters.

In this modeling approach, the shrinkage results directly from the assumption that the means are random effects from a shared distribution. Bayesians say that the means $\{\mu_i\}$ are assumed *exchangeable*, which according to a theorem by De Finetti justifies modeling them as independent conditional on a set of parameters. In my view, this formulation avoids the paradoxical nature of Stein shrinkage, because if the exchangeable assumption is unreasonable, so is the shrinkage implied for the model.

In particular, in Example 7.1, it seems reasonable (absent additional information about characteristics of the baseball players) to treat their batting averages arising from a shared distribution. In Example 7.2, it seems entirely unreasonable to assume that the percentage of foreign cars in Chicago are exchangeable with the baseball player averages. In Example 7.3, assuming exchangeability for the means from unrelated consulting problems seems completely unreasonable, and so is the idea of shrinking the estimates for each problem to the overall mean.

A standard classical analysis based on the fixed-effects model (7.3) tests the global null hypothesis $H_0: \mu_1 = \mu_2 = ... = \mu_k$ against the alternative that two or more of these means differ, using an F-test based on the ratio of between-group to within-group mean squares. If H_0 is not rejected, then differences in means are not reported out; otherwise, differences are estimated by difference in the corresponding sample means. In the random-effects model analysis, the degree of shrinkage of the group means is controlled by the size of the F statistic – the smaller the F statistic, the greater the degree of shrinkage. This analysis is more graduated than the fixed-effects analysis, which is "all or nothing."

The additional assumption that the means come from a shared distribution is not innocuous and is necessary for estimation under (7.5) to be sensible. One way of weakening the assumption is to condition on available covariate information z_i, as in the model

$$(y_{ij} \mid z_i, \mu_i, \alpha, \beta, \delta^2, \tau^2) \sim N(\mu_i, \delta^2), i = 1, ..., k, j = 1, ..., n_i$$
$$(\mu_i \mid z_i, \alpha, \beta, \delta^2, \tau^2) \sim N(\alpha + \beta z_i, \tau^2), \tag{7.7}$$

where β is an additional parameter with the same dimension as the number of covariates.

In Example 7.2, there is an obvious covariate that is ignored in the model (7.5), namely

$$z_i = \begin{cases} 1, & \text{if } i \text{ is a baseball player} \\ 0, & \text{if } i \text{ represents a percentage of foreign cars in Chicago.} \end{cases}$$

Estimates from this model stratify shrinkage on the covariate and avoid Stein's paradox, because the percentage of foreign cars and the overall baseball averages are then estimated by their own means.

Efron and Morris (1987) do not recommend including the cars with baseball players, although I find the explanation less transparent than simply considering the plausibility of the model. They write:

> Estimating the true mean for an isolated city by Stein's method creates serious errors when that mean has an atypical value. The rationale of the method is to reduce the overall risk by assuming that the true means are more similar to one another than the observed data. That assumption can degrade the

estimation of a genuinely atypical mean. Now we see why imported cars should not be included in the same calculations with the 18 baseball players. There is a substantial probability that the automobiles will be atypical.

Efron and Morris (1987) also discuss shrinkage arising from the Bayesian version of the random-effects model (7.5), writing:

> The formula for the James-Stein estimator is strikingly similar to that of Bayes's equation. Indeed, as the number of means being averaged grows very large, the two equations become identical. The James-Stein procedure, however, has one important advantage over Bayes's method. The James-Stein estimator can be employed without knowledge of the prior distribution; indeed, one need not even suppose the means being estimated are normally distributed.

I disagree; as suggested above, treating the means as coming from the common prior distribution is an inherent feature that justifies the shrinkage, and ignoring this modeling assumption allows for Stein's paradox for Example 7.2, a consequence that I seek to avoid.

7.2.3 Bayesian Decision Theory

Bayesian forms of shrinkage do not require normal errors, as in (7.5) or (7.7); other common models include the binomial model for binary outcomes, with a beta prior distribution for the component probabilities, or a Poisson model for count data, with a gamma prior distribution for the component means. Also in the Bayes world, the output of an analysis is the posterior distribution of each of the component parameters, and how that posterior distribution is summarized depends on the choice of loss function. Shen and Louis (1998) write:

> The beauty of the Bayesian approach is its ability to structure complicated models, inferential goals and analyses. The prior and likelihood produce the full joint posterior distribution which generates all inferences…. To take full advantage of the structure, methods should be linked to an inferential goal via a loss function.

In particular, the posterior mean is the best summary under quadratic loss, but the set if posterior means $\{E(\mu_i \mid \text{data})\}$ is not a good estimate of the distribution (i.e., the histogram) of component parameters, tending to "overshrink" and yield a histogram with too little dispersion.

Shen and Louis (1998) discuss alternative loss functions for the generic two-stage hierarchical model:

$$y_i \mid \theta_i \sim_{\text{ind}} l(y_i \mid \theta_i)$$
$$\theta_i \sim_{\text{iid}} G,$$

where G is a continuous prior distribution. The posterior distribution of $\theta = (\theta_1,....,\theta_k)$ is

$$g(\theta \mid Y) = l(Y \mid \theta)g(\theta)/f(Y), \text{ where } f(Y) = \int l(Y \mid s)g(s)\,ds.$$

For example, often there is interest in ranking the components of θ. The ranks of the components can be written as

$$R = (R_1,...,R_k), \text{ where } R_i = \text{rank}(\theta_i) = \sum_{j=1}^{k} I_{[\theta_i > \theta_j]},$$

where $I()$ is the indicator function. Let Q be a candidate vector of ranks. The Q that minimizes the loss function $\sum_{i=1}^{k}(Q_i - R_i)^2/k$ has components:

$$Q_j = \bar{R}_j = E(R_j \mid Y) = \sum_{i=1}^{k} \Pr(\theta_i > \theta_j \mid Y),$$

which can be made into integers by setting $\hat{R}_i = \text{rank}(\bar{R}_i)$ (Laird & Louis 1989). Shen and Louis (1998) propose compromise "triple goal" estimates that yield good estimates of the distribution of the parameters, the ranks, and the individual components.

7.3 Frequentist and Bayesian Perspectives on What Is Meant by "Random"

One of the reasons I like the Bayesian paradigm is that I have never liked the frequentist explanation of when parameters should be regarded as "fixed," as in the fixed-effects ANOVA model (7.3), and when they should be regarded as "random," as in the random-effects models (7.4) or (7.6). Here, for example, is a frequentist explanation of the difference, from Afifi and Azen (1979) (in their terminology "Model 1" is the fixed-effects model and "Model 2" is the random-effects model):

> A factor is said to be Model 1 if the particular subpopulations represented by the levels of the factor are those of interest to the experimenter. That is, if the experiment were repeated, random samples from these same subpopulations would be analyzed... In contrast, a factor is said to be Model 2 if the subpopulations represented by the levels of a factor are selected from a very large (infinite) number of subpopulations. In other words, if the experiment were repeated, it would most likely that random samples from different subpopulations would be analyzed.

Concerning this explanation:

1. It seems to me that μ_i a fixed quantity, whether or not it is sampled from a population.
2. If n_i is small, and the random-effects model is sensible, it seems that we should shrink $\bar{y}_i$ whether or not the factors are sampled from a superpopulation (Model 1).
3. On the other hand if $k = 3$, and n_i is large, I think Model 1 is appropriate, whether or not the factor is sampled from a superpopulation (Model 2). We only have two degrees of freedom for estimating τ^2, so the weight w_i is very poorly determined and shrinkage seems unnecessary.

The Bayesian perspective of what is "fixed" and what is "random" is more straightforward. Parameters or data are either "known" or "unknown." All unknowns are treated as "random" (assigned distributions) to allow quantification of uncertainty. When an unknown is observed it becomes "fixed" at the observed value and is no longer "random."

The difference between random and fixed-effects ANOVA models lies in the choice of prior distribution. The fixed-effects model (7.3) can be viewed as the limit of random effects model (7.5) as $\tau^2 \to \infty$. This "noninformative" prior implies that the between variance is so large that for a sampled subpopulation, data from other sampled subpopulations are not useful for "borrowing strength."

Setting $\tau^2 = \infty$, as in the Model 1, has the advantage that we do not need to estimate τ^2 from the data. This is useful when k is small, and information about τ^2 is limited. It is less useful when k is large, and we can get a good estimate of τ^2. I like Model 2 more when k is large enough to estimate τ^2. Exchangeability is an assumption, but Model 2 has the self-correcting feature that shrinkage is minimal when τ^2 is large relative to σ^2.

7.4 Rubin's Application

Rubin's (1980) application concerns the prediction of first-year grade average (FYA, Y) from Law School Aptitude Test (LSAT) score (X_1) and undergraduate grade point average (GPA, X_2), based on data in years 1973, 1974, and 1975 for admitted students from a set of law students. The model assumes that y_{ij}, the FYA for student j in law school i, is related the student's LSAT and GPA by the linear regression model

$$y_{ij} \sim_{\text{ind}} N(\beta_{i0} + \beta_{i1}x_{1ij} + \beta_{i2}x_{2ij}, \sigma_i^2).$$

The target quantities are $\mu_i = \beta_{i2}/\beta_{i1}$, the weight assigned to GPA relative to LSAT for school i. Rubin applies a random-effects model of the form (7.5) that targets these ratios directly, namely:

$$(a_i \mid \mu_i) \sim_{\text{ind}} N(\mu_i, v_i),$$
$$\mu_i \sim_{\text{ind}} N(\mu, \tau^2),$$

(7.8)

where $a_i = \arctan(b_{i2}/b_{i1})$ and b_{i1}, b_{i2} are the least squares estimates of β_{i1}, β_{i2}, respectively. The within-school variance of a_i, v_i is estimated by a large-sample Taylor series approximation applied to the covariance matrix of (b_{i1}, b_{i2}). The parameters (μ, τ^2) are estimated by ML, using an EM algorithm (Dempster, Laird & Rubin 1975). The performance of empirical Bayes estimates of $\{\mu_i\}$, least squares estimates of $\{\mu_i\}$, and the null method that assumes no differences in the weights across schools, is assessed by predicting GPAs in the succeeding year. The empirical Bayes estimates are much better than the null method and somewhat better than the least squares estimates, though gains are minor. The shrinkage from the random effects model is illustrated for the 1973 year data in Figure 7.2, which is Display 1 in Rubin (1980).

	Least Squares	Empirical Bayes
.0	8	
.1		
.2		
.3		
.4		
.5	8	
.6	579	
.7	3566899	1
.8	445679	399
.9	00112222233	011223334
.9	56667789999	5555666666667888889999
1.0	0011112223	00000001111111122222334
1.0	55799	556668888889999
1.1	0112345556788999	0111222233
1.2	01233447	0
1.3	8	
1.4	2	
1.5		
1.6		
1.7	0	

FIGURE 7.2

Stem and leaf plots of arctan (ratio) for 1973, estimated by least squares and empirical bayes. (Reproduced from Rubin 1980, Display 1. Copyright © 2012 American Statistical Association, reprinted by permission of Informa UK Limited, trading as Taylor & Francis Group, www. tandfonline.com on behalf of 2012 American Statistical Association.)

Rubin (1980) considered expanding the model (7.7) to include covariates z_i as in model (7.6), namely the average LSAT and UGPA scores of each school, but did not do so because strong relationships were not found.

Discussants of the paper were generally positive, with the exception of Scott (1980). A number of discussants noted the selection bias inherent in assessing predictions on admitted students, although as Rubin (1980) notes, they did not have concrete suggestions for alternatives. Dempster (1980) noted that there is "no such thing as the empirical Bayes estimator," and that the results are model-dependent. He suggested that including other school-level characteristics such as location, size, and public/private status as covariates might have improved the prediction of $\{\mu_i\}$. Scott (1980) noted the weak predictive power of LSAT and UGPA in the validation analysis. She also criticized the fact that empirical Bayes changed the ordering of school weighting relative to the least square estimates. In his rejoinder, Rubin writes:

> Scott's claim that the changing relative positions of the law schools is a counterindication for the use of empirical Bayes is unfounded. Such behavior is standard in examples with varying standard errors (e.g., see Efron & Morris 1975); estimates with large standard errors are often more extreme than estimates with small standard errors, and so should be pulled more towards a central value.

7.5 Conclusion

The formulation of random and mixed-effects models represented a giant advance in the modeling toolkit over fixed-effects models. Treating parameters as random effects leads to shrinkage of effect sizes towards zero and is a useful modeling strategy in many applied settings. The basic random effects model (7.3) can be elaborated in many ways; see, for example, Gelman and Hill (2006) on regression using multilevel models. These models play a central role in many of the topics in this book, including alternatives to least squares in multiple regression (Chapter 8), alleviating effects of multiple comparisons (Chapter 9), models for repeated measures (Chapter 10), models for multistage sample designs (Chapter 13), and models for meta-analysis of clinical trials (Chapter 14). With reference to Example 7.1, the use of statistical models in baseball, a field known as "sabermetrics," is thriving. In particular, Jensen, McShane and Wyner (2009) present a Bayesian hierarchical model that improves predictions by including prior hitting performance, age, and fielding position as covariates.

Shrinkage should be based on a plausible model that takes into account – that is, conditions on – important structure. As an example, George et al. (2017) observes that Medicare's hospital ranking system (Hospital Compare) shrinks all small hospitals to the national average AMI mortality, even though, if data from small hospitals are pooled, the mortality rate is much worse than the national average. That small hospitals perform more poorly makes sense, given that they handle a lower volume of cases, so it makes sense to include size of hospital as a covariate in the shrinkage model. Bayesians and frequentists can agree that shrinkage helps only if the model is fairly accurate, whereas Stein's paradox is often mistakenly quoted as if shrinkage is always beneficial (George et al., 2017).

7.6 Some Thought Questions on This Chapter

1. The sample mean is the best unbiased estimate of the mean for a random sample from a normal population, in terms of mean squared error loss. James-Stein showed that the sample mean is not always admissible, and the (biased) James-Stein estimator can be uniformly superior. Reconcile these two statements.

2. Consider the one-way random effects ANOVA model, with n_i observations (y_{ij}) in group $i, i = 1, ..., k$, namely:

$$(y_{ij} \mid \mu_i, \sigma^2, \tau^2) \sim N(\mu_i, \sigma^2), i = 1, ..., k, j = 1, ..., n_i,$$
$$(\mu_i \mid \sigma^2, \tau^2) \sim N(\mu, \tau^2).$$

   Derive the restricted ML (or Bayes) estimate of the mean μ_j in group j, for known μ, σ^2, τ^2? (In practice, these parameters are replaced by estimates.) What is the estimate of the mean μ_i for the fixed-effects model where μ_i is treated as a fixed parameter? Connect these results with the ideas in Efron and Morris (1977).

3. A colleague argues against the random effects model because it imposes additional assumptions on the group means, which may not be valid. Consider the pros and cons of this position in the context of (a) multisite clinical trials where the groups are study sites and (b) comparative trials of treatments for a disease where the groups consist of the alternative treatments.

4. It is argued in Efron and Morris (1977) that the James-Stein result also applies when the proportion of foreign-made automobiles is included with the baseball player averages. Is this a good idea? Explain your answer.

5. Scott (1980) in her discussion of Rubin (1980) argues that shrinkage estimates are questionable because they change the ordering of the law school estimates. Do you think this concern is appropriate?

8

Alternatives to Least Squares in Regression

The paper:

Dempster, A. P., Schatzoff, M. & Wermuth, N. (1977). A simulation study of alternatives to ordinary least squares (with discussion). *J. Amer. Statist. Assoc.*, 72, 357, 77–106.

Other readings:

Chaibub Neto, E., Bare, J. C. & Margolin, A. A. (2014). Simulation studies as designed experiments: the comparison of penalized regression models in the "large p, small n" setting. *PLoS ONE*, 9, 10, e107957.

Tibshirani, R. (1996). Regression shrinkage and selection via the Lasso. *J. Roy. Statist. Soc. Ser. B*, 58, 1, 267–288.

8.1 Introduction

Regression in its many forms is a major topic in statistics, and anyone who receives any statistical training learns about linear regression, with parameters estimated by the method of least squares. Dempster, Schatzoff and Wermuth (1977), henceforth abbreviated to DSW, features a wide-ranging simulation study of alternatives to least squares. It has been called the "57 varieties paper," because 57 forms of regression are compared. (The Heinz company introduced the "57 varieties" slogan when marketing pickles in 1896, according to Wikipedia.) DSW compares a wide variety of approaches to regression, a central topic of great interest. They frame regression in an insightful manner, and their conclusions are thought-provoking, with a hint of controversy; the controversy is perhaps more minor than in some other chapters, but alternative approaches to regression is a topic of clear applied importance, and this paper casts a lot of light on the subject.

I like good simulation studies, even though they do not have a high status in mathematical statistics compared with asymptotic theory. Asymptotic properties are important, but the sample size needed to reach "asymptotia" is unclear and varies by problem. Also, increasing the sample size while holding the model fixed has always seemed to me

DOI: 10.1201/9781003395164-10

artificial, because larger sample sizes invite more complex models with more parameters. I like well-designed simulation studies that indicate what might happen with realistic sample sizes. All simulation studies have limitations, including the one in DSW. But the simulation design here is ingenious and instructive, and it manages to cover a wide range of pertinent conditions.

Since DSW, the range of alternatives to least squares in regression has expanded considerably, and there are many choices for other readings. The paper by Tibshirani (1996) introduces another regression method, the Lasso, which had not been invented at the time of DSW. It is highly cited, and the Lasso has become of popular form of "regularized regression." It is similar in some respects to its older cousin, Ridge regression, but with some advantages if selection of covariates is of interest. The simulation study in Tibshirani (1996) is more limited, but there is no denying the impact of the paper on regression analysis. Chaibub Neto, Bare and Margolin (2014) includes an interesting simulation study comparing Ridge, Lasso, and Elastic Net, another more modern regression technique.

The holy grail of a single alternative to least squares that works in all situations is unattainable, but the search is worthwhile. Clearly, the field of statistics has advanced considerably from regression based on a linear additive model with normal errors.

8.2 The DSW Simulation Study

8.2.1 The Linear Regression Model

DSW concerns the normal linear regression model

$$(y_i \mid x_i, \beta, \sigma^2) \sim_{\text{ind}} G\left(\sum_{j=1}^{p} x_{ij}\beta_j, \sigma^2\right), \tag{8.1}$$

where $G(\mu, \sigma^2)$ is the normal distribution with mean μ and variance σ^2, $x_i = (x_{i1}, x_{i2}, ..., x_{ip})$ is a $(p \times 1)$ vector of covariates for observation i, $i = 1, ..., n$, $\beta = (\beta_1, ..., \beta_p)$ is a vector of regression coefficients, and σ^2 is the residual variance. If $x_{i1} = 1$ for all i, then β_1 is the intercept. The ordinary least squares (OLS) estimate of β is

$$b = (X^T X)^{-1} X^T y, \tag{8.2}$$

where X is the $(n \times p)$ design matrix and $y = (y_1, ..., y_n)$ is the vector of outcomes. OLS regression goes back to Gauss in the early 19th century, and the

estimates of the regression coefficients are maximum likelihood for the model
(8.1); restricted maximum likelihood captures the degrees of freedom correction when estimating the residual variance. DSW assume that the regression
model is correctly specified, so robustness to model misspecification is not
the issue. Rather, the problem addressed by DSW is over-parametrization
of the model when the sample size n small compared with the number of
covariates p, particularly when covariates are highly correlated.

8.2.2 The 57 Varieties

DSW group the alternatives to OLS into continuous and discrete shrinkage
methods.

(a) **Continuous shrinkage methods** replace the OLS estimate b by an
estimate of the form

$$\hat{\beta} = (X^T X + kQ)^{-1} X^T y,$$

where Q is a positive definite symmetric matrix and k is a nonnega-
tive scalar parameter. The choice $Q = I$, the identity matrix, leads
to Ridge regression, and the choice $Q = X^T X$ leads to a method that
DSW call Stein regression. These can be viewed as generalizations
to regression of the shrinkage estimates discussed in Chapter 7. By
applying a principal component representation of $X^T X$, DSW show
that Stein regression shrinks all components equally, whereas Ridge
"applies more drastic shrinkage where it has a greater effect in reducing mean squared error." Different versions of Ridge and Stein are
obtained by alternative approaches to estimating k. This formulation
also shrinks the intercept if a constant term is included, and in practice, it might be better to confine shrinkage to the other coefficients.

(b) **Discrete shrinkage methods** set a subset of the regression coefficients to zero, possibly after a transformation of the covariates. DSW
divide these methods into three groups:

Group 1 consists of

(b1) forward selection methods, which add at each step the covariate
that achieves the largest reduction in residual sum of squares,
stopping when the reduction is too small to be judged worthwhile according to some criterion;

(b2) backward selection methods, which start with all covariates
included, and eliminate at each step the covariate that leads
to the smallest reduction in residual sum of squares, stopping
when the reduction is judged too worthwhile to omit the covariate according to some criterion;

(b3) "Mallows CP (MCP) methods," which fit all 2^p candidate regression models and choose the one that fits best according to the CP criterion.

Group 2 procedures include

(b4) a method for determining the number of covariates r with non-zero coefficients, and then weights least squares estimates of all possible models with r covariates by their posterior probabilities (REGF). A version that includes forward selection is also considered to reduce the potentially heavy computational burden.

Group 3 includes

(b5) principal component regression (PCR) methods that transform the covariates to principle components and then set coefficients corresponding to the principal components with smallest eigenvalues to zero. This approach addresses problems with collinearity in the covariates.

Various versions of these methods make up the 57 varieties included in the simulation. I focus here on five methods, which DSW chose as representatives of the classes, based on performance and the likelihood of being used in practice. These are as follows:

OFSL: A forward selection method, where variables are included if the F statistic to enter is significant at the 5% level.

PRIF: PCR, where a principal component is included if its F statistic is significant at the 5% level.

STEINM, RIDGM: Versions of Stein and Ridge where the parameter k is chosen by a form of method of moments applied to the prior variance.

FREGF: A version of REGF where the choice of the number of covariates r is based on an F statistic.

8.2.3 The Simulation Design

DSW carried out two simulation experiments with $p = 6$ variables and $n = 20$ observations. The first experiment involved drawing one sample from each of 32 different models, obtained by fully factorial design varying five two-level factors, labeled EIG, ROT, COL, CEN, and BET. The first three factors defined the design matrix of a model. The factor EIG sets the eigenvalues of the principal component analysis of the design matrix, and took two values, namely (32, 25, 16, 9, 4, 1) and (64, 16, 4, 1, 0.25, 0.0625), setting the degree of correlation among the independent variables to be high or very high. The two levels of ROT correspond to two random replications of Gram-Schmidt orthogonal rotation matrices, with eigenvalues fixed by EIG. The third factor COL controlled the multicollinearity between individual covariates;

specifically, one level of COL introduced a correlation of 0.99 between the first two independent variables. BET and CEN jointly define β and σ^2. CEN sets two levels of the noncentrality parameter, that is, the signal to noise ratio of the model. BET sets two vectors of regression coefficients, namely (32, 25, 16, 9, 4, 1) and (64, 16, 4, 1, 0.25, 0.0625).

Experiment 2 sampled one dataset for each of 128 models, which had less extreme levels of collinearity and more variants of the correlation structure and regression coefficients. This experiment was a quarter replicate of a factorial design with five factors:

EIG with two levels, namely (30, 30, 30, 20, 20, 20) and (64, 16, 4, 2, 1, .5).

ROT with four levels, corresponding to four random orthogonal rotation matrices.

COL with two levels, whether or not a correlation of 0.95 was induced between X_1 and X_2.

MCL with two levels, whether or not a correlation of 0.92 was induced between $X_1 - X_2$ and X_3.

BET with four levels, namely (1, 1, 1, 1, 1, 1), (1, 1, 1, 0, 0, 0), (32, 16, 8, 8, 8, 8) and (32, 16, 8, 0, 0, 0).

The performance of the methods was assessed using summary metrics:

SEB = sum of squared errors of betas, a summary of how well the methods estimates the regression coefficients, and

SPE = sum of squared prediction errors, a summary of how well the methods perform for predicting the outcome.

8.2.4 Results

Table 8.1 summarizes the performance of ordinary least squares (OREG) and the five representative methods, 0FSL, PRIF, STEINM, RIDGM, and FREGF. Means of SEB and SPE are expressed as a percentage of the mean for OREG, so small is better, and values over 100 indicate inferiority to OREG. Bear in mind that Experiment 2 covers four times as many cases as Experiment 1, and in that sense should receive more weight.

It is clear from Table 8.1 that RIDGM is the winner in Experiment 2 and does relatively well in both experiments. FREGF is the winner in Experiment 1 but does less well in Experiment 2, with average SPE greater than that for OREG. Gains over OREG are much greater for regression coefficients than for predictions, particularly in Experiment 1, where the collinearity in the covariates is extreme. The PCR method, PRIF, does rather poorly for estimating regression coefficients in Experiment 1, and for prediction in both experiments; I note that PCR addresses collinearity in the predictors but does not factor in their

TABLE 8.1

Summary Performance of Six Selected Methods

Method	Experiment 1				Experiment 2			
	SEB		SPE		SEB		SPE	
	Mean	Rank	Mean	Rank	Mean	Rank	Mean	Rank
OREG	100	57	100	27	100	56	100	21
OFSL	15	22	84	16	72	30	123	38
PRIF	19	31	162	52	35	5	120	37
STEINM	89	54	104	28	81	41	93	9
RIDGM	8	1	87	22	22	2	80	1
FREGF	9	2	73	6	55	14	119	35

Source: Reproduced from Dempster, Schatzoff and Wermuth (1977), Table 8.1, copyright © 2012 American Statistical Association, by permission of Informa UK Limited, trading as Taylor & Francis Group, www.tandfonline.com on behalf of 2012 American Statistical Association.

Note: Mean SEB and Mean SPE expressed as percentage of respective means for ordinary least squares regression (OREG).

relationship to the outcome, accounting for its mediocre performance for prediction. STEINM shows minor gains over OREG relative to RIDGM.

DSW also include an analysis of the factors that are varied in the simulation experiments. Their summary conclusions are:

> Substantial improvements over OREG are exhibited when collinearity effects are present, noncentrality in the original model is small, and selected true regression coefficients are small. Ridge regression emerges as an important tool, while a Bayesian extension of variable selection proves valuable when the true regression coefficients vary widely in importance.
>
> The potential relative gains in accuracy for individual estimated regression coefficients are seen to be typically much larger than those for predicted Y values. The expected gains in SEB and SPE are fairly small for STEINM, while for RIDGM and PRIF they are larger and exhibit dependencies ... mainly on eigenvalues and noncentrality factors. Methods such as OFSL or FREGF, which select variables, are highly dependent additionally on the pattern of true regression coefficients. The relatively conservative STEINM procedure has one advantage, in that it less often does worse than OREG. The relative accuracy of traditional selection of variables methods based on significance testing behaves erratically in relation to significance levels adopted.

8.2.5 Limitations

All simulation studies have limitations, and this one is no exception; bear in mind that the computing power in the 1970s was a small fraction of what is available now. The DSW experiments lack any replication within each simulation condition and are focused solely on mean squared error of estimates.

Specifically, no attention is given to statistical inference, for example, by assessing width and confidence coverage of interval estimates. Because the model underlying OREG is correct, one would expect t confidence intervals for the regression coefficients and prediction intervals for Y to have nominal coverage, although the intervals will tend to be excessively wide. I would expect Bayesian versions of Ridge and Stein to improve on RIDGM and STEINM because the prior distribution can reflect uncertainty in the parameter k. Stepwise methods like forward or backward selection use the data twice, to determine the choice of model and inference under the model, and this can be expected to have a negative effect on statistical inferences.

8.2.6 The Discussion of DSW

An interesting feature of the discussion of DSW, pointed out by the authors in their rejoinder, is that it mainly concerns the comparison of Ridge and Stein. Largely ignored were stepwise methods and PCR, which were much more common than Ridge and Stein as alternatives to least squares at the time DSW was written and are still common in applications to this day.

The main "controversy" in the discussion concerned whether the superiority of Ridge was real or an artifact of the simulation design. Thus, the thrust of the discussions by Thisted, Efron and Morris, and Bingham and Larntz is that "it all depends on the model." The superior performance of Ridge over Stein is a result of the simulation choices, and other simulation designs could have produced different results. In particular from a Bayesian perspective, Ridge and Stein correspond to different prior distributions for the regression coefficients. The fact that Ridge beats Stein in this simulation must reflect the fact that the set of simulation conditions maps more closely into the Ridge prior than the Stein prior.

DSW do not dispute this point, and indeed emphasize it in the main paper, writing (their Equation [2.1] is Equation [8.2] in this chapter):

> The general estimator (2.1) has a simple Bayesian interpretation. If β has the multivariate normal prior distribution $N(0, \omega^2 Q^{-1})$ then the posterior distribution of β is $N(\hat{\beta}, (X^T X / \sigma^2 + Q / \omega^2)^{-1}$, where $\hat{\beta}$ is determined by (2.1) with k given by $k = \sigma^2 / \omega^2$. Thus $\hat{\beta}_R$ is a posterior mean corresponding to a prior $N(0, \omega^2 I)$ distribution for β, and $\hat{\beta}_S$ is a posterior mean corresponding to a prior $N(0, \omega^2 (X^T X)^{-1})$ distribution for β.

The Ridge prior in effect assumes exchangeability of the regression coefficients, in the same way that the random effects model in Chapter 8 assumed exchangeability in the means. Whether this is a reasonable assumption depends on the application, although I confess I find the prior underlying Ridge to be more plausible than the prior underlying Stein, and Ridge regression methods (frequentist or Bayesian) play a much more prominent role in subsequent applications.

DSW in their discussion focus on empirical Bayes versions of Ridge and Stein and note that the effectiveness of a given method depends on how the shrinkage parameter k is estimated. They consider a simple method of moments estimator, but alternatives are to compute maximum likelihood estimates, or to apply full Bayes by assuming a prior distribution for k.

8.3 Later Work

One might think that 57 alternatives to least squares are enough, but many more new methods have been developed since DSW, and this remains a very active area of research. The Lasso (Tibshirani 1996) is a popular alternative to Ridge. This penalized regression method replaces the quadratic (L2 norm) penalty associated with Ridge, which minimizes

$$\sum_{i=1}^{n}(y_i - X_i\beta)^2 + \lambda\sum_{j=1}^{p}\beta_j^2, \tag{8.3}$$

by the L1 norm penalty, which minimizes

$$\sum_{i=1}^{n}(y_i - X_i\beta)^2 + \lambda\sum_{j=1}^{p}|\beta_j|. \tag{8.4}$$

The fitting criterion has the property that it tends to shrink small regression coefficients to zero, making it more readily amenable than Ridge to variable selection (although note that estimating an effect to be zero is not the same as saying the true effect is zero). Simulations in Tibshirani (1996) suggest that the Lasso predictions are better than Ridge when the true model has a sparse number of non-zero coefficients, whereas Ridge outperforms Lasso when the true model has a substantial number of small to moderate sized regression coefficients. However, the simulation studies in this paper do not systematically vary factors in the way that DSW attempts, and the results do not lend themselves to general conclusions.

Chaibub Neto, Bare and Margolin (2014) compare Ridge and Lasso with another method, Elastic Net (Zou & Hastie 2005), a hybrid of Lasso and Ridge that penalizes least squares by a convex combination of L2 and L1 norms. To their credit, these authors argue for applying principles of good experimental design into the design of simulations and make an attempt to capture the parameter space. They focus on prediction for "small n large p" applications where the number of covariates is potentially much greater than the number of observations. They summarize their conclusions as follows:

> Comparison of Ridge regression against Lasso corroborates two well-known results, namely: (i) that also outperforms Ridge when the true model is sparse, whereas the converse holds true for saturated models... and (ii) for highly correlated features, Ridge tends to dominate Lasso in terms of predictive performance... Comparison of Ridge regression against Elastic Net corroborates the well known result that Elastic Net tends to show much better performance than Ridge regression when the true model is sparse, while these methods tend to be comparable for saturated models.

Frequentists call shrinkage methods that penalize the fitting criterion, as in Equations (8.3) and (8.4), "regularization." The alternative, more prominent in DSW, is to interpret penalization as in effect assuming a prior distribution on the regression coefficients. Under the Bayesian interpretation, the prior distribution elucidates conditions under which the corresponding shrinkage is justified; it also justifies a fully Bayesian analysis where a prior distribution is assumed for the penalty parameter λ in (8.3) or (8.4). This analysis propagates uncertainty in the parameter, which is treated as known in a frequentist analysis. In the context of the Lasso, for example, see the Bayesian Lasso (Park & Casella 2008).

Both Tibshirani (1996) and Chaibub Neto, Bare and Margolin (2014) adopt the regularization interpretation of shrinkage, although Tibshirani (1996) notes the interpretation of the Lasso as a Bayesian method with a double exponential prior for the regression coefficients. These authors focus on mean squared errors of predictions and do not consider statistical inference, where putting a prior distribution on the penalty parameter may lead to an improvement in estimating standard errors. Neither of these papers refer to DSW and do not consider that the performance of the methods depends on the suitability of the prior, as brought out in the DSW's more Bayesian discussion. Indeed, Chaibun Neto, Bare and Margolin (2014) view the compared methods as algorithms, with very little discussion of the underlying statistical models. This is a pity, from my point of view. Although Tibshirani (1996) and Chaibub, Bare and Margolin (2014) concern methods other than Ridge not included in DSW, the review of regression and philosophy of their simulations is instructive.

For me, these alternatives to least squares for the normal regression model, and extensions to non-normal regression models, are best viewed as adding informative prior distributions for the regression coefficients. Many recent methods fall under this general framework, including Bayesian additive regression trees (BART, Chipman, George & McCullough 2010), regression with horseshoe priors (Carvalho, Polson & Scott 2010), regression with spike and slab priors for variable selection (George & McCulloch, 1997), and Bayesian logistic regression (Gelman et al. 2008).

8.4 Some Thought Questions on This Chapter

1. What are in your view the main take-home conclusions from the DSW paper and discussion. Is there anything that surprised you? Something new that you were not aware of before reading it?

2. What are the main strengths and weaknesses of the DSW simulation design?

3. What's the main difference between Ridge and the Lasso method in Tibshirani? Discuss when the Lasso might be a superior approach.

4. What are the Bayesian and frequentist approaches to least squares alternatives? Discuss pros and cons of these perspectives.

5. A reader of the chapter wrote: with the 57 varieties in DSW along with later developments, how does one decide which to use in a $n < p$ setting in practice? Discuss.

9

Multiple Perspectives on Multiple Comparisons

The papers:

Benjamini, Y. & Hochberg, Y. (1995). Controlling the false discovery rate: a practical and powerful approach to multiple testing. *J. Roy. Statist. Soc. Ser. B,* 57, 1, 289–300.

Berry, D. (2012). Multiplicities in cancer research: ubiquities and necessary evils. *J. Nat. Cancer Inst.,* 104, 1124–1132.

Cox, D. R. (1965). A remark on multiple comparison methods. *Technometrics,* 7, 2, 223–224.

Gelman, A., Hill, J. & Yajima, M. (2012). Why we (usually) don't have to worry about multiple comparisons. *J. Res. Educ. Effectiveness,* 5, 189–211.

Rothman, K. J. (1990). No adjustments are needed for multiple comparisons. *Epidemiology,* 1, 43–46.

Tukey, J. W. (1991). The philosophy of multiple comparisons. *Statist. Sci.,* 6, 1, 100–116.

9.1 Introduction

The question of whether and when multiple comparison corrections are needed for tests or confidence intervals is one of the more fascinating controversies in statistical inference. Distinguished statisticians line up on both sides of the issue. Tukey's (1991) paper puts the argument in favor of corrections and presents an important example of such a method. Many books have been written about multiple comparison corrections; the Bonferroni method, which modifies the size of a test by dividing it by the number of comparisons under consideration, is the most frequently applied, perhaps because of its simplicity. Benjamini and Hochberg (1995) propose an alternative to the more conventional corrections that controls the false discovery rate (FDR), the expected proportion of falsely rejected hypotheses. I discuss this approach in Section 9.5. On the other side, Cox (1965), Gelman, Hill and Yajima (2012) and Rothman (1990) basically adopt the position that multiple comparison corrections are usually unnecessary, Berry (2012) provides a nuanced discussion, which I'll quote extensively in the concluding section.

DOI: 10.1201/9781003395164-11

The Wikipedia entry on the "Multiple Comparisons Problem" states:

> In statistics, the multiple comparisons, multiplicity or multiple testing problem occurs when one considers a set of statistical inferences simultaneously or estimates a subset of parameters selected based on the observed values. The larger the number of inferences made, the more likely erroneous inferences become. Several statistical techniques have been developed to address that problem, typically by requiring a stricter significance threshold for individual comparisons, so as to compensate for the number of inferences being made.

For example, if 10 independent tests are carried out at the nominal level of 0.05, and all the null hypotheses are true, then the probability that at least one of the tests is rejected is $(1-0.95^{10}) = 0.40$ – there is a 40% chance of getting at least one false-positive, which represents the "erroneous inference" in the above quote.

The simple and ubiquitous Bonferroni correction divides the nominal level by the number of tests, in this case giving a multiple comparison-corrected nominal level of $0.05/10 = 0.005$. At this level, the probability that at least one of the tests is rejected is $(1-0.995^{10}) = 0.049$, close to the original nominal level of 0.05. If the tests are not independent, the Bonferroni correction tends to be conservative, motivating other correction methods.

Some argue that multiple comparison corrections are unnecessary. If many questions are tested without adjustment, then the number of false positives increases, but the validity of any one, considered in isolation, is not affected. Rothman (1990) writes:

> The purported problem with all these "significant" P-values is that many null hypotheses will be rejected even if they are correct. Of course, there is nothing peculiar about conducting a multitude of comparisons, as opposed to a single comparison, that increases the probability of rejecting a specific null hypothesis when it is correct. If $\alpha = 0.05$, there is a five percent probability of rejecting a correct null hypothesis, whether one or one billion are examined. The core of the supposed problem is that with many comparisons the number of potentially incorrect statements regarding null hypotheses will be large, simply because of the large number of comparisons.

The following quote from Cox (1965) is often cited by those arguing against multiple comparison adjustments:

> The essence of the matter seems to me to be this. The fact that a probability can be calculated for the simultaneous correctness of a large number of statements does not usually make that probability relevant for the measurement of the uncertainty of one of the statements.

Three facets of the issue cast light on the question of whether multiple comparison corrections are needed:

a. Comparison-wide versus experiment-wide error, and which is more appropriate for the inference;
b. Selection effects when particularly significant comparisons are highlighted, and how to account for them in the inference;
c. A clear statement of the problem; some of the confusion arises from ambiguity about how the question is posed.

I take up these points in the rest of the chapter.

9.2 Comparison-wise versus Experiment-Wise Error Rates

In an analysis that involves more than one comparison, should the error rate of a test refer to false rejection of any of the comparisons – the experiment-wise error rate – or false rejection of each individual comparison – the *comparison-wise* error rate. The Bonferroni-type correction is applied to control the former, but no correction is needed for the latter. Tukey (1991) adopts the experiment-wise perspective in the following quote:

> In many situations we could look at any or all of many things, and would if the amount of available data were not severely restricted. One of these is clinical trials - of drugs or therapies - where both ethical and financial considerations limit the size of the trials. In such situations, it is often essential to focus sharply on only one or two primary questions, questions that deserve analysis in terms of confident directions (and which often may as well be analyzed in terms of confidence intervals). We can then spend all (if there is one question) or half (if there are two) of our permissible error rate, often 5% overall, on the single primary question, or on the two primary questions.
>
> Once we have spent this error rate, it is gone. And what we say about the remaining questions has to have many of the properties of hints, even if we work at, say, an individual error rate of 5%. Keeping the different strength with which we believe primary and auxiliary answers – especially when these answers appear to use the same statistics (e.g., Student's t) in the same way – is a very serious and important challenge. The message has to be that it can be wise and necessary to focus on a very few prespecified questions, prespecified before data collection, whenever we cannot enjoy the luxury of enough data to work with either familywise (F) or Bonferroni (B) error rates.

The advice to prespecify a small number of primary questions seems wise, whether or not an adjustment for multiple comparisons is contemplated. Other hypotheses should be carefully labeled as exploratory, requiring validation from an independent study if there is evidence of a signal.

Three common situations leading to multiple hypotheses are considered in Examples 9.1–9.3: (a) comparisons of multiple treatments or conditions; (b) comparisons of treatments or conditions in subsets of the population; and (c) comparisons of treatments on multiple outcome measures.

Example 9.1. Comparisons of multiple treatments

Consider the question of the comparative effectiveness of two drugs A and B, as measured by the difference in the sample means of some primary outcome. Consider two possible designs: (a) an independent groups design comparing drugs A and B; and (b) an independent groups design comparing A and B and three other treatments, say C, D, and E. Sample sizes per treatment group are the same in (a) and (b). In (a), there is one comparison of interest, so no multiple comparison issue. In (b), there are 10 distinct pairwise comparisons. Should the inference comparing A and B be adjusted for the 10 possible comparisons?

One way of reducing the multiple comparison problem is to confine attention to a smaller number of contrasts. For example, with two new treatments corresponding to two doses and one control treatment, instead of considering the three pairwise differences, compare the average of the new treatments against control, and the difference in the new treatments.

However, Rothman's (1990) argument, quoted above, suggests that a multiple comparison correction is not needed in comparisons of multiple treatments. If designs (a) and (b) got the same data regarding the comparison of A and B, why should the inferences be different? The availability of additional data for treatments C, D, and E in design (b) might affect the specifics of the analysis – for example, the data might be included in a pooled estimate of the variance, or a random-effects model might be assumed to borrow strength if the individual treatment groups are based on small samples. But the nominal level for tests or confidence intervals should not be changed because design (b) allows for more treatment comparisons.

If, however, the question is not the comparison of A and B, but rather the treatment that yields the largest treatment effect, then some allowance for selection is needed, as discussed in Section 9.3. Failure to distinguish between these questions is one reason for divergent views on whether multiple comparison adjustments are needed.

Example 9.2. Subset comparisons

Treatment comparisons in subsets of the population are undeniably of interest. In the clinical trial context, the term "precision medicine" is predicated on the idea of targeting treatments according to patient characteristics, requiring an assessment of treatment effects in subsets defined by those characteristics. If a substantial number of subset comparisons are carried out and a few are statistically significant in comparison-wide tests, the argument for making some allowance for multiple comparisons seems strong, because selection of subsets with large treatment effects is implicit:

1. Reporting only significant subset comparisons without specifying the full set of comparisons tested is faulty, because the selection of significant results alters the frequentist P-values. A related issue is regression to the mean, in that large estimated treatment effects are liable to be biased upward because of selection, as discussed in Chapter 7.
2. Listing all the comparisons made is an improvement over (1), but evaluating individual significant comparisons as part of a set seems difficult. For example, if the overall treatment effect is not significant, and say two out of 20 comparisons made are significant at the 5% level, the evidence for any substantive differences seems very tenuous at best, particularly given that many studies are powered for the overall treatment effect and have low power for assessing subgroup comparisons.

Berry (2012) writes the following concerning this issue:

> Regression to the mean is pronounced when examining subsets, at least in part because there are many types of patient characteristics (eg, disease stage, age, diet, race, geography) and consequently many possible subsets … and opportunities to be wrong… Consistent with regression to the mean, the vast majority of subset analyses that are taken seriously by sponsors and investigators fail to be confirmed and so lead to disappointment. Part of the tantalizing aspect of subsets is psychological, with the victims grasping for a positive straw in a negative haystack…. The fundamental dilemma of modern cancer research is that we cannot make progress without examining effects within subsets. Only a minority of patients benefit from conventional cancer therapies. Administering every available therapy to every patient is not tenable. We must learn which patients benefit and which do not—a subset problem. So we cannot dismiss subset analyses simply because they are dangerous and usually wrong. Proper methodology will entail some level of confirmation. We walk on the thinnest of ice, but we know the ice is thin and so we walk gingerly … or run!

That is, significant subset comparisons need to be considered with a large grain of salt, and validation on independent data seems crucial to avoid being misled.

Example 9.3. Multiple outcomes

When multiple comparisons arise from multiple outcome measures, the arguments for a multiple comparison correction seem to me weaker. If the outcomes are of a different nature, such as outcomes measuring treatment benefit and other outcomes measuring side effects, then the experiment-wise error rate seems inappropriate, because the consequences of significant treatment effects are quite different for the two types of measures. If the outcomes are reflecting different aspects of a shared underlying treatment effect, then summarizing them in a single index seems preferable to adjusting each one for multiple comparisons. Cook and Farewell (1996) write:

> In clinical trial designs formally based on two or more responses, multiplicity adjustments may not be necessary if marginal, or separate, test results are *interpreted marginally* and have implications in *different aspects* of the prescription of the treatments (i.e. response specific effects are of interest and separate statements regarding them are desired). Thus there may be contexts in which multiple tests of significance should be performed with reference to marginal rather than experimental error rates. If hypothesis tests are primarily directed at marginal inferences, it is then reasonable to specify a maximum tolerable error rate for each specific hypothesis test. This point can be highlighted by careful consideration of the precise nature of the possible type I errors and the meaningfulness of an 'overall' type I error, i.e. at the design stage of a clinical trial an investigator will have to consider the nature and relative frequency of the type I errors that will be tolerated.

Cook and Farewell (1996) consider as alternatives to Bonferroni-corrected marginal comparisons tests of a global null hypothesis of no treatment effect for all the outcome measures, and tests based on summary statistics, such as summing the ranks of treatments on the outcomes. Their preferred solution varies depending on context.

9.3 Accounting for the Effects of Selective Inference

Some form of multiple comparison correction is necessary if the inference is subject to selection. In the words of Cox (1965):

> Suppose that we have samples of equal size from k populations with means $\mu_1, ... \mu_k$, that the usual normal-theory assumptions are made and that make separate confidence limit statements about all the $\mu_i - \mu_j$, using the simple t multiplier. So long as we look at the statements individually, an arbitrarily selected one satisfies the requirements for a confidence limit. A probability of error referring to the simultaneous correctness of a set of statements seems relevant only if a certain conclusion depends directly on the simultaneous correctness of a set of the individual statements. The central difficulty enters when we look not at all $\mu_i - \mu_j$, but only at some, selected partly in the light of the data. (It might be better to talk about the problem of selected comparisons rather than about the problem of multiple comparisons).

Berry (2012) dramatizes the selection issue by constructing a bogus prognostic marker for breast cancer:

> I built a one-dimensional prognostic score based on a subset of markers that were prognostic for relapse-free survival. Twenty biomarkers were available for each of the 1550 patients in the trial. I selected five markers from these 20 because they had P values less than .10 in univariate proportional hazards regression. Then I built a multivariable regression equation, combining the five markers into a single score, with higher scores indicating worse prognosis.

The resulting marker seemed to be prognostic of breast cancer, with a hazard ratio for breast cancer of 0.77 for individuals in the sample below vs. above the median value of the marker ($P < 0.0001$). The good performance was entirely spurious and the result of selection bias, because all of the markers were created from random noise and had no prognostic power whatsoever. For guidance on appropriate methods for creating biomarkers, Berry references McShane et al. (2005).

When assessing a set of comparisons, the largest estimated comparison might be of particular interest, as in the following example.

Example 9.4. Inference for the largest difference versus inference for a particular difference

Consider again a clinical trial comparing six treatments for a medical condition. Treatment effects might be measured by the difference in the mean of an outcome in each pair of treatments. Should multiple comparison corrections be applied in this situation? There are 15 distinct pairwise comparisons, but only five of these are "distinct," in that the other 10 can be derived by combining the five chosen comparisons. If one of the six treatments is a control treatment, one might consider the five comparisons between the control and the other treatments. A simple Bonferroni correction would thus divide the nominal level by five, replacing 0.05 by 0.01.

If a particular comparison arose was randomly selected from the set of five comparisons with control, there seems little need for a correction. On the other hand, if the particular comparison was picked because it was the largest, then the selection effect needs to be taken into account. A frequentist approach for normal outcome data is Tukey's studentized range test (Tukey 1991), which is based on the distribution of the maximum of the five differences in means. A Bayesian approach is to compute the posterior distribution of the maximum of the differences in means. This distribution is readily computed by drawing five differences from their posterior distributions, and then selecting the maximum difference from these draws. The posterior probability that each new treatments is the best is also readily computed. See, for example, Gelman et al. (2013, Section 5.5).

9.4 Bayesian Hierarchical Modeling

Gelman, Hill and Yajima (2012), and Berry in his earlier work, argue that the Bayesian approach to inference largely addresses the issue of multiple comparisons. Gelman, Hill and Yajima (2012) write:

> Our approach, as described in this article, has two key differences from the classical perspective. First, we are typically not terribly concerned with Type 1 error because we rarely believe that it is possible for the null hypothesis to be strictly true. Second, we believe that the problem is not multiple testing but rather insufficient modeling of the relationship between the corresponding parameters of the model. Once we work within a Bayesian multilevel modeling framework and model these phenomena appropriately, we are actually able to get more reliable point estimates. A multilevel model shifts point estimates and their corresponding intervals toward each other (by a process often referred to as "shrinkage" or "partial pooling"), whereas classical procedures typically keep the point estimates stationary, adjusting for multiple comparisons by making the intervals wider (or, equivalently, adjusting the p values corresponding to intervals of fixed width). In this way, multilevel estimates make comparisons appropriately more conservative, in the sense that intervals for comparisons are more likely to include zero. As a result we can say with confidence that those comparisons made with multilevel estimates are more likely to be valid. At the same time this "adjustment" does not sap our power to detect true differences as many traditional methods do.

Specifically, Bayesian hierarchical modeling addresses the problem of regression to the mean discussed in Section 9.2, because the extremes of a set of multiple comparisons are pulled toward the mean, by treating the parameters that model treatment effects as random. This reduces the problem of selection of largest estimated effects, while not completely solving it.

A common frequentist approach to the multiple comparison issue is to perform a test of the global null hypothesis that all the treatment effects are zero, as in the F test from the fixed effect ANOVA model discussed in Chapter 7. If the global null hypothesis is not rejected, then no individual treatment differences are reported. If the global null hypothesis is rejected, then individual treatment effects are tested. A Bonferroni-type correction is often applied to the individual treatment effects, although this seems overkill, in that this correction for multiple comparisons renders the F test screen unnecessary.

As noted in Chapter 7, the shrinkage estimates from a Bayesian mixed model analysis are related to the F test screen, with the F statistic in the F test determining the degree of shrinkage of the individual comparisons. The Bayesian approach seems to me more satisfying, in that the F test screen is "all or nothing," but effects are rarely exactly zero. The Bayesian approach compromises between treating all comparisons as zero and estimating the comparisons at their individual point estimates.

I find the argument for Bayesian hierarchical models convincing, but do not think it completely addresses the issue of selection of significant effects from a set of multiple comparisons. The one-way random effects model involves exchangeability assumptions that may be unrealistic, and information for estimating between group variances is limited if the number of comparisons is small. The potential bias from selection is reduced but not eliminated. However, Bayesian inference methods for selected parameters are available, as discussed in Example 9.4.

9.5 The False Discovery Rate

The feature of multiple comparison corrections often criticized by critics is that the global null that all the compared comparisons are zero is unreasonable. For example, in a genome-wide association study, the expectation is that a small number of genes are causing the disease. The job is to find these genes, not test that none of the genes have any effect.

Benjamini and Hochberg (1995) addresses this concern by controlling the FDR, the expected proportion of rejected null hypotheses that are in fact true. Suppose m null hypotheses are tested simultaneously, of which m_0 are true. Of the m_0 that are true, suppose u are declared non-significant and v are declared significant. Of the $m - m_0$ that are false, suppose t are declared non-significant and s are declared significant (see Table 9.1). The family-wise error rate (FWER) is then $\Pr(v \geq 1)$, and the per comparison error rate (PCER) is $E(v/m)$. The proportion of hypotheses that are falsely declared significant is $q = v/(s + v)$, so the FDR is defined as $E(q) = E(v/s + v)$.

It is easily shown that when all null hypotheses are true, that is, $m_0 = m$, the FDR is equivalent to the FWER, and when $m_0 < m$, the FDR is smaller than or equal to the FWER, so the FDR is less stringent, and a gain in power may be expected.

TABLE 9.1

Errors Committed When Testing a Set of Null Hypotheses

	Declared Non-significant	Declared Significant	Total
True null hypothesis	u	v	m_0
Non-true null hypothesis	t	s	$m - m_0$
Total	$m - r$	r	m

Source: Benjamini and Hochberg (1995), Table 1. Reproduced with permission of John Wiley & Sons through PLSClear.

Benjamini and Hochberg (1995) describe the following simple procedure for controlling the FDR at a particular value q^*:

> Consider testing H_1, H_2, ..., H_m based on the corresponding p-values P_1, P_2, ..., P_m. Let $P_{(1)} \leq P_{(2)} \leq ... \leq P_{(m)}$ be the ordered p-values, and denote by $H_{(i)}$ the null hypothesis corresponding to $P_{(i)}$. Define the following Bonferroni-type multiple-testing procedure: let k be the largest i for which $P_{(i)} \leq (i/m)q^*$; then reject all $H_{(i)}$, $i = 1, 2, ..., k$.
>
> *Theorem 1.* For independent test statistics and for any configuration of false null hypotheses, the above procedure controls the FDR at q^*.

The fact that the FDR makes good scientific sense, and the availability of a simple procedure for achieving it, has made the Benjamini and Hochberg (1995) paper justly popular. Bayesian versions of the FDR have also been developed; see, for example, Storey (2003).

9.6 Key Points and Recommendations From Berry (2012)

I summarize by presenting some key points and recommendations from Berry (2012), a nuanced discussion of the issue. He notes that multiple comparisons present a complicated problem, and there is no "panacea, no one-size-fits-all approach." Some key points from his paper follow:

Multiplicities matter. "Multiple observations increase the probability of false-positive conclusions and have led to many false and otherwise misleading publications ....The number of comparisons is critical in this development. But few investigators report denominators for their comparisons and few journals insist on it...."

Issues in publication. "Ignoring multiplicities is implicit in the traditional approach to publication when different comparisons are addressed in the same experiment and published separately. It makes no sense to draw a different conclusion about comparison X when comparison Y is also made than when only comparison X is made ... and comparison Y is made in another publication!"

Silent multiplicities. "Especially problematic are "silent multiplicities," observations and potential observations that have not been reported by the researcher and whose existence the researcher may not even recognize .... A major distinction of the latter is that assessing the denominator of possibilities may be impossible .... Keeping silent through the publication process has become conventional in academia, and it is encouraged by journals' attitudes regarding what research is publishable. Reports of research may be the truth but not the whole truth. What readers have not been told may be more important than what they have been told."

Implications for exploratory analysis. "Teaching exploratory analyses, say, without recognizing multiplicities and their impact is a disservice to students and to science, and it can lead them to deceive the rest of us."

Serendipitous results. "Given the many opportunities to observe something unusual, an apparently extraordinary observation becomes quite ordinary."

Selection bias. Berry's bogus biomarker study produces a phony prognostic index purely from noise.

Regression to the mean "is ubiquitous and powerful in medical research and in science generally. Failing to appreciate its impact has led many researchers astray. An example is "the placebo effect." When patients improve after taking a placebo, many researchers attribute the improvement to the placebo when in fact it is almost always due to regression to the mean."

Bayesian analysis. "Arguments against adjusting for multiple comparisons have a Bayesian flavor. It is well known that the Bayesian approach 'takes the data at face value' when calculating statistical measures of evidence. It is based on the probability of the data actually observed, as compared with frequentist P values that include probabilities of tail areas and of other data that might have occurred but did not …. However, the Bayesian approach is not as cavalier as 'take the data at face value' may suggest …. In the Bayesian approach, the background information of the trial is part of the prior probability distribution of the various parameters. The number of comparisons affects the prior distribution, but assessing a prior distribution in the context of multiplicities is complicated."

Berry (2012) concludes with some recommendations for investigators and journals:

1. *Need for a detailed protocol.*

2. *Indicate analyses in the protocol that were not done.*

3. *Keep a log of actual analyses, including those not specified in the protocol.*

4. *Unspecified analyses. Discoveries from analyses not specified in the protocol may be publishable, but the relationship between the goal of the study and the discovery should be made clear in the publication, including whether the discovery was serendipitous.*

5. *Adjusting for multiplicities is usually preferable, not essential if 1 through 3 above are satisfied.*

6. *Confirmation is an essential aspect of discovery.*

7. *Piecemeal publication. The full study and previous publications and analyses (including those not published) should be described.*

8. *It is best to have and to give a biological explanation for any empirical observation, especially if the explanation occurred before the observation. The human mind is capable of retrospectively building a rationale for any observation, including when the observation is wrong!*

9. *Frequentist versus Bayesian. Both approaches are acceptable. The philosophical differences lead to some differences in handling multiplicities in the inferential process, but the basic principles of the need to "adjust" or "shrink" unusual observations are similar. And both have similar attitudes to confirmation and the advantages of prospective studies.*

10. *REMARK criteria. For biomarker studies, follow the REMARK criteria* (McShane et al. 2005) *in addition to the above.*

9.7 Some Thought Questions on This Chapter

1. A large randomized study compares five alternative treatments, yielding 10 possible distinct comparisons between treatments. Consider the following four approaches to analysis:

   a. Compute *P*-values for the appropriate test at the 5% level for all comparisons, with no multiple comparison adjustment.

   b. Apply a Bonferroni correction, computing the tests in (a) and dividing the nominal alpha level for the tests by 10.

   c. Test the global null hypothesis at 5% level; if the test is not rejected, report no differences. If the test is rejected, do (a) above.

   d. Test the global null hypothesis at 5% level; if the test is not rejected, report no differences. If the test is rejected, do (b) above.
      State which of the procedures (a–d) you prefer, and briefly justify your choice.

2. In a randomized study as in 1, suppose we are comparing just two treatments, but are comparing them on 10 distinct outcomes, so the 10 comparisons are for different outcomes for the same two treatments. Which if the four options in 1 do you favor? Again, briefly justify your choice.

3. With large numbers of comparisons, shrinkage methods (Bayes or empirical Bayes) have been suggested as a solution to the multiple comparison problem. Do you agree? Discuss briefly.

10

Generalized Estimating Equations

The paper:

Liang, K.-Y. & Zeger, S. L. (1986). Longitudinal data analysis using generalized linear models. *Biometrika*, 73, 1, 13–22.

Other readings:

Hubbard, A. E., Ahern, J., Fleischer, N. L., Van Der Laan, M., Lippman, S. A., Jewell, N., Bruckner, T. & Satariano, W. A. (2010). To GEE or not to GEE. Comparing population average and mixed models for estimating the associations between neighborhood risk factors and health. *Epidemiology*, 21, 467–474.

Pepe, M. S. & Anderson, G. L. (1994). A cautionary note on inference for marginal regression models with longitudinal data and general correlated response data. *Commun. Statist. – Simul. Comp.*, 23, 4, 939–951.

10.1 Introduction

A major undercurrent in the history of statistics is the debate between "full probability modeling," with the dual inferential pillars of maximum likelihood (ML) and Bayes, and what might be termed "algorithmic modeling," where the statistical model is not necessarily fully specified, and inference is based on algorithms or estimating equations. Thus, in Chapter 1, we discussed Fisher's famous (1922a) paper, which proposed ML as an alternative to the method of moments as a general principle of estimation. Advocates of algorithmic modeling argue that the approach provides fast computation and robustness to model misspecification. Full probability modelers argue that the model lays bare the assumptions underlying the inference, which are then capable of criticism and modification.

An important development in algorithmic modeling was generalized estimating equations (GEEs), which extend the range of algorithmic modeling far beyond the method of moments applied to Pearson's system of distributions, as discussed in Fisher's (1922a) paper. Perhaps the most practically important application of GEE was introduced in the famous paper, which is the topic of this chapter. Specifically, Liang and Zeger (1986), henceforth LZ, described GEE for a class of models for longitudinal and clustered data, a

 DOI: 10.1201/9781003395164-12

setting where full probability models were then largely confined to the case of multivariate normal outcomes.

In other reading, Pepe and Anderson (1994) discuss an important assumption about the covariates in the GEE model that was implicit in LZ. Hubbard et al. (2010) compare the assumptions in GEE and full probability models for repeated-measures data with nonlinear link functions. The instructive points raised by these papers are discussed in Sections 10.2 and 10.3 below, but before considering GEE and these commentaries, Examples 10.1 and 10.2 describe two major earlier applications of the full probability modeling approach that set the stage for the LZ paper.

Example 10.1. The GLIM model

A major advance in applied statistics in the 1970s was the software package Generalized Linear Interactive Modeling (GLIM), developed by a working group of the Royal Statistical Society (Baker & Nelder 1983; McCullagh & Nelder 1989; Nelder 1975). This software package provided ML inference for a very useful class of regression models, extending multiple linear regression models to a variety of non-normal outcomes.

Suppose the data consists of n units $(y_i, x_{i1}, ..., x_{ip})$, $i = 1, ..., n$, of an outcome variable Y and p predictor variables $X_1, ... X_p$. Given $x_i = (x_{i1}, ..., x_{ip})$, the n values of y_i are assumed to be an independent sample from a regular exponential family distribution of the form

$$f(y_i \mid x_i, \beta, \phi) = \exp\left[\left(y_i\theta_i - a(\theta_i) + b(y_i)\right)\phi\right], \qquad (10.1)$$

where $\theta_i = h(\eta_i), \eta_i = x_i\beta$, and h is a known link function. The first two moments of y_i given x_i are

$$E(y_i \mid x_i, \beta, \phi) = a_i'(\theta_i), \ \ \text{Var}(y_i \mid x_i, \beta, \phi) = a_i''(\theta_i)/\phi,$$

where $a_i'(\theta_i)$ and $a_i''(\theta_i)$ are the first and second derivatives of $a_i(\theta_i)$ with respect to θ_i. Important special cases of this model are:

Normal linear regression: $h = $ identity, $a(\theta) = \theta^2/2, \phi = 1/\sigma^2$;
Poisson regression: $h = \log, a(\theta) = \exp(\theta), \phi = 1$;
Logistic regression: $h = $ logit, where $\text{logit}(\mu_i) = \log\left(\mu_i/(1 - \mu_i)\right)$, $a(\theta) = \log(1 + \exp(\theta)), \phi = 1$

The loglikelihood of $\theta = (\beta, \phi)$ given observed data (y_i, x_i), $i = 1, ..., n$ based on the model (10.1) is

$$\ell_Y(\beta, \phi \mid y) = \sum_{i=1}^{n}\left[\left(y_i\theta_i - a(\theta_i) + b(y_i)\right)\phi\right],$$

which for non-normal cases does not generally have an explicit maximum. Numerical maximization can be achieved using an iterative algorithm, such as Fisher scoring (e.g. McCullagh & Nelder 1989, Section 2.5). In particular, the score equations for β take the form

$$U(\beta) = \sum_{i=1}^{n} \delta_i x_i^T (y_i - a_i'(\theta_i)) = 0, \text{ where } \delta_i = \frac{d\theta_i}{d\eta_i}. \quad (10.2)$$

The model (10.1) does not include models for longitudinal or clustered data where the observations for a given unit or cluster i cannot realistically be assumed independent. Models for this type of data in the 1970s were largely confined to normal outcomes, as in the following model.

Example 10.2. Normal repeated-measures model

In longitudinal studies, units are observed at different times and/ or under different experimental conditions. The following general repeated-measures model is given in Jennrich and Schluchter (1986) and builds on earlier work by Hartley and Rao (1967), Laird and Ware (1982) and others. Suppose that the hypothetical complete data for unit i consist of K measurements $y_i = (y_{i1}, \ldots, y_{iK})$ of an outcome variable Y, and that y_i are independent with distribution

$$(y_i \mid X_i, \beta, \phi) \sim_{\text{ind}} N_K(X_i\beta, \Sigma(\phi)), \quad (10.3)$$

where X_i is a known $(K \times m)$ design matrix for unit i, β is a $(m \times 1)$ vector of unknown regression coefficients, and the elements of the covariance matrix $\Sigma(\phi)$ are known functions of q unknown parameters ϕ. The model thus incorporates a mean structure, defined by the set of design matrices $\{X_i\}$, and a covariance structure, defined by the form of the covariance matrix $\Sigma.(\phi)$. The observed data consist of the design matrices $\{X_i\}$ and outcome measurements $\{y_i, i = 1, \ldots, n\}$.

A large number of practical situations can be modeled by combining different choices of mean and covariance structures. Common covariance structures include:

Independence: $\Sigma = \text{Diag}_K(\phi_1, \ldots, \phi_K)$, a diagonal $(K \times K)$ matrix;

Compound symmetry: $\Sigma = \phi_1 U_K + \phi_2 I_K$, ϕ_1 and ϕ_2 scalar, $U_K = (K \times K)$ matrix of ones, $I_K = (K \times K)$ identity matrix;

Autoregressive, lag 1: $\Sigma = (\sigma_{jk})$, $\sigma_{jk} = \phi_1 \phi_2^{|j-k|}$, ϕ_1, ϕ_2 scalars;

Banded: $\Sigma = (\sigma_{jk})$, $\sigma_{jk} = \phi_{rj-k}$, where $r = |j-k|+1, r = 1, \ldots, K$;

Factor-analytic: $\Sigma = \Gamma\Gamma^T + \phi_0$, with $\Gamma = (K \times q)$ the matrix of "factor loadings", $\phi_0 = (K \times K)$ the diagonal matrix of "specific variances," and $\phi = (\Gamma, \psi_0)$;

Random Effects: $\Sigma = Z\phi^* Z^T + \sigma^2 I_K$, where Z is a $(K \times q)$ known matrix, ϕ^* is a $(q \times q)$ dispersion matrix, σ^2 is a scalar, I_K the $K \times K$ identity matrix, and $\phi = (\phi^*, \sigma^2)$;

Unstructured: $\Sigma = (\sigma_{jk})$, with ψ representing the $v = K(K+1)/2$ elements of this matrix.

The mean structure is also very flexible. If $X_i = I_K$, the $(K \times K)$ identity matrix, then $\mu_i = \beta^T$ for all i. Between-subject and within-subject effects are readily modeled through other choices of X_i.

The loglikelihood for the model (10.3) is

$$\ell_Y(\beta, \phi) = -0.5n \log|\Sigma(\phi)| - 0.5 \sum_{i=1}^{n} (y_i - X_i\beta)^T \Sigma^{-1}(\phi)(y_i - X_i\beta), \quad (10.4)$$

which is linear in the observed quantities $\{y_i, y_i^T y_i, i = 1, \ldots, n\}$. This likelihood does not yield explicit ML estimates except in simple special cases. For the unstructured covariance structure $\Sigma(\phi) = \Sigma$, explicit ML estimates are available for β given Σ, and for Σ given β. Hence, an iterative ML solution is obtained by the following "alternating conditional modes" algorithm: Given estimates $(\beta^{(t)}, \Sigma^{(t)})$ at iteration t, estimates at iteration $t + 1$ are

$$\beta^{(t+1)} = \sum_{i=1}^{n} \left(X_i^T \left(\Sigma^{(t)}\right)^{-1} X_i \right)^{-1} X_i^T \left(\Sigma^{(t)}\right)^{-1} y_i, \quad (10.5)$$

$$\Sigma^{(t+1)} = \frac{1}{n} \sum_{i=1}^{n} \left(y_i - X_i^T \beta^{(t+1)} \right)^T \left(y_i - X_i^T \beta^{(t+1)} \right). \quad (10.6)$$

ML estimation is also feasible when there are missing values in the repeated measures y_i, assuming the missingness mechanism is missing at random (MAR; see, e.g., Little and Rubin [2019], chapter 1). An alternative to ML is to add a prior distribution for the parameters and base inferences on their posterior distribution. Computation is facilitated using Markov Chain Monte Carlo methods, as discussed in Chapter 11.

The goal of Liang and Zeger (1986) (LZ) was to generalize the models in Examples 10.1 and 10.2 to longitudinal or clustered data with non-normal outcomes. Their method is computationally relatively straightforward, although in general iterative. The principle of estimation was via GEEs rather than ML, and the model specified a mean and "working" covariance structure but does not require a full probability model for the joint distribution of the outcome variables. The approach also has a form of robustness, in that the working covariance structure used to model dependence between the repeated measures does not have to be correctly specified to get valid estimates and estimates of uncertainty. For these reasons, GEE has remained a popular approach to this day.

After LZ, methods were developed that do ML and Bayes for correlated non-normal data, through generalized linear mixed models (e.g., PROC NLMIXED in SAS.) Some comments comparing the ML and GEE approaches are provided in Section 10.3.

10.2 GEE for Longitudinal Data

We now replace the scalar outcome y_i in Example 10.1 with a vector $y_i = (y_{i1}, ..., y_{iK})$ and $x_i = (x_{i1}, ..., x_{iK})$, where y_{it} is the outcome and x_{it} is the set of covariates for unit i at time t. LZ assume the marginal density of y_{it} given x_i has the same form as Eq. (10.1) in Example (10.1), that is:

$$f(y_{it} \mid x_i, \beta, \phi) = \exp\left[\left(y_{it}\theta_{it} - a(\theta_{it}) + b(y_{it})\right)\phi\right], \tag{10.7}$$

where $\theta_{it} = h(\eta_{it})$, $\eta_{it} = x_{it}\beta$. One approach to estimation is to ignore any dependence within repeated measures and simply assume an independence correlation structure and apply the methods of Example 10.1. The resulting score equation for β is

$$U(\beta) = \sum_{i=1}^{n} X_i^T \Delta_i S_i = 0, \tag{10.8}$$

where $\Delta_i = \text{Diag}(d\theta_{it}/d\eta_{it})$ is a $(K \times K)$ diagonal matrix and $S_i = y_i - a_i'(\theta_i)$ is a $(K \times 1)$ vector for unit i. LZ note that the solution of this equation gives consistent estimates of the parameters even when the repeated measures are correlated. That is, the ML estimates for the GLIM under independence remain consistent even though the correlation structure of the repeated measures is being misspecified. Valid large-sample standard errors can be computed using a sandwich estimator, or by bootstrapping (see Chapter 11), leaving out the complete set of repeated measures for each unit rather than just the individual values.

These estimates are consistent but no longer efficient if the repeated measures are correlated. The efficiency of these estimates can then be improved by replacing the diagonal matrix Δ_i in Eq. (10.8) by a matrix that incorporates a realistic working correlation structure. Specifically, replace (10.8) by

$$U(\beta) = \sum_{i=1}^{n} D_i^T V_i^{-1} S_i = 0, \tag{10.9}$$

where $D_i = d\{a_i'(\theta)\}/d\beta = A_i\Delta_i X_i$, and $V_i = A_i^{1/2}R(\alpha)A_i^{1/2}$, where $R(\alpha)$ is a working correlation structure indexed by an $(s \times 1)$ parameter vector α. Provided ϕ is estimated consistently, for example, by the method of moments, estimates of β based on solving (10.9) are again consistent whether or not the working correlation structure is correctly specified, and are more efficient than estimates based on (10.8) when the working correlation structure $R(\alpha)$ is closer to the truth.

Some comments on this approach follow:

a. **The normal case.** For the normal model with h the identity function and $\phi = 1/\sigma^2$, Eq. (10.9) is score equation for β, and GEE reduces to ML if ϕ is also estimated by maximizing the likelihood. In this case, the robustness property of GEE under misspecification of the covariance structure is shared by ML. GEE is not more robust than ML in this setting, because they are one and the same method, at least for estimation of β given ϕ.

b. **Time-varying covariates.** If the covariates x_i are time-varying, that is x_{it} depends on t, then the model assumes that

$$f(y_{it} \mid x_i, \beta, \phi) = f(y_{it} \mid x_{it}, \beta, \phi); \qquad (10.10)$$

that is, the conditional distribution of the outcome at time t, y_{it}, given x_i depends only on the values of the covariates x_{it} at time t. A model that conditions on the full set of covariates x_i conditions on values of past and future covariates for unit i, in addition to the values of the covariates at time t. A *cross-sectional* model for y_{it} at time t concerns the distribution of y_{it} given x_{it}, but the correlation of y_{is} and y_{it} is only well defined if it conditions on a common set of covariates x_i.

The assumption (10.10) is implicit in LZ, but it is easy to miss because the notation in the paper does not always specify the values of X on which distributions are being conditioned.

As an example of being clear about conditioning, consider a simple normal one-way random effects model, where y_{ij} is the value of an outcome variable for unit j within group i. A common way of writing this model is:

$$y_{ij} = \mu_i + \varepsilon_{ij}, \ \mu_i \sim N(\mu, \tau^2), \ \varepsilon_{ij} \sim N(0, \sigma^2).$$

A Bayesian form of the model, with weak prior distributions, is:

$$\left[y_{ij} \mid \mu_i, \mu, \sigma^2, \tau^2 \right] \sim_{\text{ind}} N(\mu_i, \tau^2)$$

$$\left[\mu_i \mid \mu, \sigma^2, \tau^2 \right] \sim_{\text{ind}} N(\mu, \tau^2)$$

$$\pi(\mu, \sigma^2, \tau^2) \propto 1/\sigma^2.$$

The notation is clear about conditioning and independence assumptions, which becomes more important as models become more complex.

Pepe and Anderson (1994) point out the assumption (10.10) that is implicit in LZ, noting that it is often unreasonable in applied settings. They write:

> To illustrate, consider an example discussed by Zeger & Liang (1992) concerning Vitamin A deficiency. The question addressed is whether

Vitamin A deficiency is associated with an increased incidence of respiratory disease, controlling for other covariate components such as weight, height, age and gender... In the Vitamin A example discussed by Zeger & Liang (1992) the correlation among outcomes $\{Y_{ij}, j = 1,..., n_i$ is likely to derive from an observation-driven process rather than from a parameter-driven process. Having a respiratory infection at one time point will directly increase the risk of being infected subsequently. The increased risk is not likely to be totally explained by a latent random variable common to all time points indicating, say, 'susceptibility to respiratory infection'. Nevertheless, Zeger & Liang (1992) propose to analyze the data with a cross-sectional model using an estimating equation (2) with non-diagonal working covariance matrix R. Our arguments in this paper, however, suggests that a diagonal matrix might be preferable in this setting since the cross-sectional model is of primary interest and condition {Eq. (10.10)} is likely not to hold.

When (10.10) does not hold, a model that conditions on future values of the covariates is often not appropriate from a causal perspective. It may make more sense to condition the model for y_{it} on current and past values of the covariates, or more generally, on the history up to time t, including prior values $\{y_{is}, x_{is}, s < t\}$. For this reason, GEE should be used with caution when the objective is regression coefficients that have a causal interpretation.

The terms "conditional," "marginal," and "cross-sectional" are common in the literature, but I confess that I am not a big fan of them. The usual distinction is that marginal models for Y do not condition on values of Y at other times points, whereas conditional models do. But marginal models condition on covariates, and hence are not truly marginal in a probabilistic sense. Also, a variety of "conditional" models might be entertained, depending on what variables are included in the conditioning. "Cross-sectional" models at time t would appear to condition only on outcomes and covariates measured at time t, but the term is often used interchangeably with "marginal models" which in the GEE sense need to condition on the full set of covariates, including values at times other than t. It is all a bit confusing, in my view.

10.3 GEE versus Generalized Linear Mixed Models

For longitudinal data with non-normal errors, estimation based on (10.9) differs from ML for a fully specified repeated-measures model. The latter approach would be fully efficient if the model was correctly specified, but potentially more vulnerable to model misspecification. LZ write:

> For non-Gaussian outcomes, however, less development has taken place. For binary data, repeated measures models in which observations for a

subject are assumed to have exchangeable correlations have been proposed by Ochi & Prentice (1984) using a probit link, by Stiratelli, Laird &
Ware (1984) using a logit link and by Koch et al. (1977) using log linear
models. Only the model proposed by Stiratelli, Laird & Ware allows for
time dependent covariates. Zeger, Liang & Self (1985) proposed a first-
order Markov chain model for binary longitudinal data which, also,
however, requires time independent covariates. One difficulty with the
analysis of non-Gaussian longitudinal data is the lack of a rich class of
models such as the multivariate Gaussian for the joint distribution of
$Y_{it}, t = 1,..., n$. Hence likelihood methods have not been available except in
the few cases mentioned above.

After the SZ paper, generalized linear mixed models (GLMMs) were developed that extend GEE models to longitudinal or clustered data. See, for
example, Breslow and Clayton (1993). As in Section 10.2, we replace the scalar
outcome y_i in Example 10.1 with a vector $y_i = (y_{i1},...., y_{iK})$ and $x_i = (x_{i1},...., x_{iK})$,
where y_{it} is the outcome and x_{it} is the set of covariates for unit i at time t. A
typical GLIMM assumes that

$$f(y_{it} \mid x_i, z_i, \delta, \gamma, \phi) = \exp\left[\left(y_{it}\theta_{it} - a(\theta_{it}) + b(y_{it})\right)\phi\right], \tag{10.11}$$

where $\theta_{it} = h(\eta_{it}), \eta_{it} = x_{it}\delta + z_{it}\gamma_i, \gamma_i \sim_{\text{ind}} N(0, \Gamma)$. Dependence of repeated measures is modeled by including random effects γ_i that are assumed to be nor-
mally distributed across units. The resulting distribution of the outcome
variable y_i is obtained by integrating out the random effects from Eq. (10.11):

$$\int f(y_i \mid x_i, z_i, \delta, \phi) f(\gamma_i \mid z_i, \gamma) dz_i = f(y_i \mid x_i, z_i, \delta, \gamma, \phi),$$

which does not necessarily have a known functional form. ML for this model
is complex, and a variety of approximate methods have been proposed (e.g.,
Breslow & Clayton 1993). The Bayesian approach is amenable using Markov
Chain Monte Carlo methods, as discussed in the next chapter. See, for example, Hadfield (2010).

For nonlinear links, the interpretation of the coefficients δ in this model
differs from the interpretation of the coefficients β in the GEE model (10.7),
because f conditions on the random effects γ_i. Consider, for example, the following example with a single covariate and binary outcomes:

Example 10.3. GEE and GLLMs for binary outcomes

Suppose now the data for unit i consist of vectors $y_i = (y_{i1},...., y_{iK})$ and
$x_i = (x_{i1},...., x_{iK})$, where at time t, y_{it} is a binary outcome taking values 0
and 1, and x_{it} is the value of a single covariate. A possible GEE model for
these data assumes the logit link function:

$$\text{logit } E(y_{iy} \mid x_i, \beta) = \beta_{0t} + \beta_{1t}x_{it}, \text{ or equivalently } E(y_{it} \mid x_i, \beta) = \frac{\exp(\beta_{0t} + \beta_{1t}x_{it})}{1 + \exp(\beta_{0t} + \beta_{1t}x_{it})}.$$

The parameter β_{1t} is then the effect of increasing x_{it} on the log odds that $y_{it} = 1$. The corresponding GLMM with compound symmetry correlation structure assumes that

$$\text{logit } E(y_{it} \mid x_i, \gamma_i, \delta) = \delta_{0t} + \delta_{1t} x_{it} + \gamma_i,$$
$$\text{or equivalently } E(y_{it} \mid x_i, \gamma_i, \delta)$$
$$= \frac{\exp(\delta_{0t} + \delta_{1t} x_{it} + \gamma_i)}{1 + \exp(\delta_{0t} + \delta_{1t} x_{it} + \gamma_i)},$$

and the parameter δ_{1t} is the effect of increasing x_{it} on the log odds that $y_{it} = 1$, given the value of the random effect γ_i. The interpretation is tricky, because the random effect is unobserved, and (unlike the interpretation of β_t), it is dependent on correct specification of the correlation structure, The inclusion of random effects γ_i within the link function is a disturbing feature, which leads some authors to prefer the GEE formulation. For example, Hubbard et al. (2010) write:

> We argue in general that mixed models involve unverifiable assumptions on the data-generating distribution, which lead to potentially misleading estimates and biased inference. We conclude that the estimation-equation approach of population average models provides a more useful approximation of the truth.

An alternative to GEE is to formulate a joint model that does not include random effects in the link function for the mean. For example, in the setting with binary outcomes, y_{it} could be assumed to be a binary variable that takes the value 1 when an underlying latent variable y_{it}^* crosses a threshold, and then $\{y_{it}^*\}$ modeled using a normal repeated-measures models such as that in Example 10.2. For example:

$$(y_i^* \mid X_i, \beta, \phi) \sim_{\text{ind}} N_K\left(X_i\beta, \Sigma(\phi)\right), \ y_i^* = (y_{i1}^*, ..., y_{iK}^*),$$
$$y_{it} = 1 \text{ if } y_{it}^* < 0, y_{it} = 0 \text{ if } y_{it}^* \geq 0, \tag{10.12}$$

where (without loss of generality) the diagonal elements of the covariance matrix $\Sigma(\phi)$ are set to one. If, in particular, it is assumed that $E(y_{it}^* \mid X_i, \beta, \phi) = \beta_{0t} + \beta_{1t} x_{it}$, then this model leads to a probit link function for y_{it}:

$$\Pr(y_{it} = 1 \mid x_i, \beta, \phi) = \Pr(y_{it}^* < 0 \mid x_i, \beta, \phi) = \Phi(\beta_{0t} + \beta_{1t} x_{it}),$$

where Φ is the cumulative normal (or probit) function. This link is perhaps harder to interpret than the logit, but it avoids including the random effects inside the link function, the problem raised by Hubbard et al. (2010) described above. The article by Ochi and Prentice (1984) cited above is a special case of this strategy. More comparisons of this probit approach with GEE under correctly specified and misspecified models would be of interest.

10.4 Missing Data

Both the GEE method of LZ based on (10.9) and likelihood methods based on GLMM's such as in Section 10.3 can accommodate missing data in the repeated measures, although the assumptions about missingness mechanisms are different. GEE in general requires that missingness is missing completely at random, in the sense that missingness can depend on the covariates (assumed fully observed) but not on the repeated measures. This is a strong assumption, often violated in practice. If missingness is in the form of attrition, where some individuals drop out before the end of the study, then weighted GEE can applied, weighting units that drop out at time t. See, for example, Preisser, Lohman and Rathouz (2002). This approach yields consistent estimates providing dropout depends on the covariates and the recorded history of the outcomes until time t, a form of missing at random (MAR, see Rubin 1976). Likelihood methods based on GLMMs make a similar MAR assumption. The likelihood analysis relies more on modeling assumptions, but avoids weighting, which can add noise to the estimates when missingness is only weakly related to the repeated measures.

For a general pattern of missing data in the repeated measures, likelihood-based analyses remain consistent under the MAR assumption that missingness does not depend on the missing values after conditioning on the observed data, recorded before or after the time point in question. This assumption is weaker than the assumptions underlying GEE, and in fact, weights are not readily defined for a general nonmonotone pattern of missing data – the simplest general approach is to weight the complete cases, which (like unweighted GEE) only allows missingness to depend on fully observed covariates.

10.5 Conclusions

The LZ approach to repeated-measures data with non-normal outcomes provides some robustness to misspecification of the correlation structure and avoids the computational complexity of GLMMs, which have parameters that are hard to interpret because they condition on unknown random effects. On the other hand, GEE inferences are asymptotic, whereas the Bayesian approach to GLMMs does not assume large samples and is computationally feasible via MCMC methods.

Concerning the issue of time-varying covariates discussed in Section 10.2, it is important to be clear about what exactly are being included as covariates in the model. The choice of conditioning needs to be carefully justified, particularly if parameters are intended to have a causal interpretation.

10.6 Some Thought Questions on This Chapter

1. "ML makes more assumptions than GEE, so GEE is therefore more robust." Discuss in the context of the models considered in this chapter.

2. Since LZ, more explicit likelihood-based methods based on GLMMs have become available. What are the relative strengths and weaknesses of GEE over these more explicitly model-based methods?

3. What exactly is meant by a "marginal model," specifically with reference to the conditioning on the covariates? Is there an implicit assumption here?

4. Compare the assumptions about the missingness mechanism made by GEE, weighted GEE, and GLMMs when some of the repeated measures are missing, for both a monotone and a general pattern of missing data.

11

The Bootstrap and Bayesian Monte-Carlo Methods

The papers:

Efron, B. (1979). Bootstrap methods: another look at the jackknife. *Ann. Statist.*, 1–26.
Gelfand, A. E. & Smith, A. F. M. (1990). Sampling-based approaches to calculating marginal densities. *J. Amer. Statist. Assoc.*, 85, 410, 398–409.

Other readings:

DiCiccio, T. J. & Efron, B. (1996). Bootstrap confidence intervals. *Statist. Sci.*, 11, 3, 189–228.
Rubin, D. B. (1981). The Bayesian Bootstrap. *Ann. Statist.* 9, 1, 130–134.
Tanner, M. A. & Wong, W. H. (1987). The calculation of posterior distributions by data augmentation. *J. Am. Statist. Assoc.*, 82, 398, 528– 540.

11.1 Introduction

The rapid increase in computational power over the last 50 years has had a profound influence on statistical analysis. In the days when Fisher (1922a) was laying out the theory of maximum likelihood (ML), statistical analysis was often based on tables of a few basic distributions like the normal, Student's t, F, and chi-squared distributions; tractable analysis was restricted to a limited class of models. Eventually, iterative methods of estimation using algorithms like scoring, Newton-Raphson, and EM extended the applicability of ML, at least if the number of parameters was not prohibitively large. However, Bayesian analysis was restricted to models with small numbers of parameters because of the high-dimensional integrations involved.

In this chapter, I discuss two widely used approaches to computation in statistical modeling, one frequentist and one Bayesian. The frequentist approach is the bootstrap (Efron 1979), a sampling approach that simulates the distribution of a statistic, treating the distribution of the sample as an approximation of the underlying distribution in the population. The approach provides frequentist inference when analytical approaches are intractable. While various extensions of the bootstrap have been developed, the original form has the advantage of simplicity and widespread applicability. In Section 11.2, I review the main features of this basic version of the bootstrap, together with the jackknife, an earlier sample reuse method for bias reduction and estimation of sampling errors.

For Bayesian inference, I discuss methods that generate draws from the posterior distribution. I focus specifically on two methods, direct simulation and the Gibbs' sampler, a Markov Chain Monte-Carlo (MCMC) method originally developed in the context of analysis of high-dimensional imaging data (e.g., Geman & Geman 1984). The key feature of these methods is that, as discussed in Section 11.3.1, they avoid the problems of multidimensional integration, which previously limited Bayesian methods to models with an analytic solution or a modest number of parameters. I choose Gelfand and Smith (1990) as the main paper, a seminal and highly cited paper that brings together a number of MCMC approaches, together with important applications to multilevel models.

In other reading, DiCiccio and Efron (1996) review methods for bootstrap confidence interval estimation that address some of the limitations of the basic bootstrap method. Rubin (1981) describes the Bayesian bootstrap, a Bayesian analog of the bootstrap based on a multinomial model with a Dirichlet prior distribution. Discussed here as Example 11.2, it provides an example of a direct simulation method for simulating the posterior distribution. Tanner and Wong (1987) provide an early and important application of the Gibbs' sampler to missing data.

11.2 The Jackknife and the Bootstrap

Efron (1979) introduced the bootstrap as a more versatile alternative to the Jackknife (Quenouille 1949), a method for reducing the bias of estimates and estimating their variance without parametric assumptions. Let $\hat{\theta} = f_n(x_1,...,x_n)$ be an estimate of a scalar parameter θ based on an independent sample $x = (x_1,...,x_n)$. Let $\hat{\theta}_{(i)} = f_{n-1}(x_1,...,x_{i-1},x_{i+1},...,x_n)$ be the estimate of θ obtained by the same estimator applied to the data omitting the ith observation. Denote the average of these estimates as $\hat{\theta}_{\text{jack}} = \sum_{i=1}^{n} \hat{\theta}_{(i)}/n$. The jackknife estimate of the bias of $\hat{\theta}$ is $\hat{b}(\hat{\theta}) = (n-1)(\hat{\theta}_{\text{jack}} - \hat{\theta})$, and the bias-corrected jackknife estimate of θ is

$$\hat{\theta} - \hat{b}(\hat{\theta}) \equiv \hat{\theta}_{\text{jack}}^{*} = n\hat{\theta} - (n-1)\hat{\theta}_{\text{jack}}. \tag{11.1}$$

A Taylor series expansion shows that if $\hat{\theta}$ has a bias of order $1/n$, then $\hat{\theta}_{\text{jack}}^{*}$ has a bias of order $1/n^2$.

Define the ith pseudo-value to be $P_i = n\hat{\theta} - (n-1)\hat{\theta}_{(i)}$, which average to $\hat{\theta}^*_{\text{jack}}$. The jackknife estimate of the variance of $\hat{\theta}$ is then

$$V_{\text{jack}}(\hat{\theta}) = s^2_{\text{jack}}/n, \text{ where } s^2_{\text{jack}} = \sum_{i=1}^{n}(P_i - \hat{\theta}^*_{\text{jack}})^2/(n-1), \qquad (11.2)$$

the sample variance of the pseudo-values.

The set of jackknife samples here are obtained by leaving out one unit – if n is large, versions can also be defined that leave a set of more than one units out. The basic version of the bootstrap replaces the "leave one out" replicates by data obtained by sampling n units *with replacement* from the set of n units in the sample. Formally, for $i = 1, \ldots, n$, let m_i denote the number of times unit i is included in the sample. Then $\sum_{i=1}^{n} m_i = n$, and $(m_1, \ldots, m_n)$ has a multinomial distribution with index n and probabilities $(1/n, \ldots, 1/n)$. Operationally, a bootstrap sample is created by drawing this set of counts from the multinomial distribution, an easily accomplished task.

Let $\{S_b, b = 1, \ldots, B\}$ denote a set of B bootstrap samples created in this way, and let $\hat{\theta}_{(b)}$ be the estimate of θ computed on the bth bootstrap sample. The bootstrap distribution of $\hat{\theta}$ is then the empirical distribution given by $\{\hat{\theta}_{(b)}, b = 1, \ldots, B\}$. If this distribution is approximately normal, a 95% confidence interval for θ is given by

$$I_{.95,\text{boot}}(\theta) = (\bar{\theta}_{\text{boot}} \pm 1.96 s_{\text{boot}}), \qquad (11.3)$$

where $\bar{\theta}_{\text{boot}} = \sum_{b=1}^{B} \hat{\theta}_{(b)}/B$ and $s^2_{\text{boot}} = \sum_{b=1}^{B}(\hat{\theta}_{(b)} - \bar{\theta}_{\text{boot}})^2/(n-1)$; the original estimate $\hat{\theta}$ can also replace $\bar{\theta}$ in this expression.

If the bootstrap distribution is not approximately normal, a better 95% confidence interval than (11.3) is

$$I_{.95,\text{boot}}(\theta) = (\hat{\theta}_{\text{boot},.025}, \hat{\theta}_{\text{boot},.975}), \qquad (11.4)$$

where the lower and upper limits in (11.4) are the 2.5th to 97.5th sample percentiles of the bootstrap distribution. A considerably larger value of B, say $B = 5000$, is needed to estimate the percentiles with sufficient accuracy. Thus, the method is computationally quite intensive. This percentile approach tends to outperform the jackknife in situations where the distribution of $\hat{\theta}$ is highly non-normal.

Efron (1979) shows that the jackknife can be derived as a Taylor series approximation of the bootstrap. The bootstrap is more versatile, in that it produces estimates of variance and confidence intervals for statistics like the median, where the jackknife fails. The jackknife can also be shown to yield more conservative intervals than the bootstrap.

The generality, conceptual simplicity, and intuitive appeal of the bootstrap has made it a very popular method. While nonparametric, in the sense that no statistical model is assumed, it is not "assumption-free." The estimate being bootstrapped needs to be consistent to get valid tests and confidence intervals. Quantities being bootstrapped need to be exchangeable ("iid") for the method to work. The theory of the nonparametric method is asymptotic and does not necessarily yield confidence intervals close to nominal levels in small samples. For example, Schenker (1985) shows that bootstrap confidence intervals for the variance of a normal distribution do not cover well for sample sizes around 100, a quite substantial number. The conditions under which the method works well are not easy to check.

As discussed elsewhere, Bayesian methods are less reliant on the assumption of large samples, and another approach to improving the small-sample properties of the bootstrap is to apply its Bayesian analog, namely the Bayesian bootstrap (Rubin 1979), which is discussed in Example 11.2.

Suppose θ is a $(p \times 1)$ vector parameter in a statistical model and $\hat{\theta}$ is the ML estimate. As discussed in Chapter 1, a standard estimate of the covariance matrix of $\hat{\theta}$ is the inverse of the observed or expected information matrix. Bootstrapping the ML estimate is an alternative way of computing the estimated covariance matrix, and it can be shown to be asymptotically equivalent to the sandwich estimate of the covariance matrix. This estimate tends to be robust to misspecification of certain features of the model, such as the variance structure in normal repeated-measures models.

Another useful feature of the bootstrap estimate of the covariance matrix of $\hat{\theta}$ is that it avoids the need to compute and invert a $(p \times p)$ observed or expected information matrix, which can be very burdensome in problems where p is large. This feature is particularly useful in missing-data problems where ML estimation is achieved using the EM algorithm (Dempster, Laird & Rubin 1977), because (unlike alternative algorithms like scoring or Newton-Raphson) EM does not invert an information matrix at each iteration. The following canonical missing-data problem (see, e.g., Little & Rubin 2019, Chapter 11) provides an example:

Example 11.1. Bootstrap estimates of variance for ML estimates for the multivariate normal data with a general pattern of missing values

Suppose $(Y_1, Y_2, \ldots, Y_K)$ have a K-variate normal distribution with mean $\mu = (\mu_1, \ldots, \mu_K)$ and covariance matrix $\Sigma = (\sigma_{jk})$. The data have a haphazard pattern of missing values, and write $Y = (Y_{(0)}, Y_{(1)})$ where Y represents a random sample of size n on $(Y_1, \ldots, Y_K)$, $Y_{(0)}$ the set of observed values, and $Y_{(1)}$ the missing data. Let $y_{(0),i}$ represent the set of variables with values observed for unit i, $i = 1, \ldots, n$. The loglikelihood based on the observed data is then:

$$\ell\left(\mu, \Sigma | Y_{(0)}\right) = \text{const} - \frac{1}{2} \sum_{i=1}^{n} \ln\left|\Sigma_{(0),i}\right| - \frac{1}{2} \sum_{i=1}^{n} \left(y_{(0),i} - \mu_{(0),i}\right)^T \Sigma_{(0),i}^{-1} \left(y_{(0),i} - \mu_{(0),i}\right), \quad (11.5)$$

where $\mu_{(0),i}$ and $\Sigma_{(0),i}$ are the mean and covariance matrix of the observed components of Y for unit i.

EM is an appealing computational approach to ML estimation for this problem. The hypothetical complete data Y belong to the regular exponential family with sufficient statistics

$$S = \left(\sum_{i=1}^{n} y_{ij}, \quad j=1,\ldots,K; \quad \sum_{i=1}^{n} y_{ij} y_{ik}, \quad j,k=1,\ldots,K \right).$$

At the tth iteration of EM, let $\theta^{(t)} = \left(\mu^{(t)}, \Sigma^{(t)}\right)$ denote current estimates of the parameters. The E step of EM for iteration $t + 1$ calculates the expected value of the complete data sufficient statistics S, conditional on the observed data and current parameter estimates:

$$E\left(\sum_{i=1}^{n} y_{ij} \mid Y_{(0)}, \theta^{(t)} \right) = \sum_{i=1}^{n} y_{ij}^{(t+1)}, \quad j=1,\ldots,K \tag{11.6}$$

and

$$E\left(\sum_{i=1}^{n} y_{ij} y_{ik} \mid Y_{(0)}, \theta^{(t)} \right) = \sum_{i=1}^{n} \left(y_{ij}^{(t+1)} y_{ik}^{(t+1)} + c_{jki}^{(t+1)} \right), \quad j,k=1,\ldots,K, \tag{11.7}$$

where

$$y_{ij}^{(t+1)} = \begin{cases} y_{ij}, & \text{if } y_{ij} \text{ is observed} \\ E(y_{ij} \mid y_{(0),i}, \theta^{(t)}), & \text{if } y_{ij} \text{ is missing;} \end{cases}$$

and

$$c_{jki}^{(t+1)} = \begin{cases} 0 & \text{if } y_{ij} \text{ or } y_{ik} \text{ is observed} \\ \text{Cov}(y_{ij}, y_{ik} \mid y_{(0),i}, \theta^{(t)}), & \text{if } y_{ij} \text{ and } y_{ik} \text{ are missing.} \end{cases}$$

Missing values y_{ij} are thus replaced by the conditional mean of y_{ij} given the set of values, $y_{(0),i}$ observed for that unit and the current estimates of the parameters, $\theta^{(t)}$. These quantities are easily computed using the sweep operator (Little & Rubin 2019, Section 7.4.3). The M step of the EM algorithm is straightforward, computing new estimates $\theta^{(t+1)}$ of the parameters from the expected complete data sufficient statistics. That is,

$$\mu_j^{(t+1)} = n^{-1} \sum_{i=1}^{n} y_{ij}^{(t+1)}, \quad j=1,\ldots,K;$$

$$\sigma_{jk}^{(t+1)} = n^{-1} E\left(\sum_{i=1}^{n} y_{ij} y_{ik} \mid Y_{(0)}, \theta^{(t)} \right) - \mu_j^{(t+1)} \mu_k^{(t+1)} \tag{11.8}$$

$$= n^{-1} \sum_{i=1}^{n} \left[(y_{ij}^{(t+1)} - \mu_j^{(t+1)})(y_{ik}^{(t+1)} - \mu_k^{(t+1)}) + c_{jki}^{(t+1)} \right], \quad j,k=1,\ldots,K.$$

EM iterates between (11.7) and (11.8) until convergence is achieved. Note that this algorithm does not require computing and inverting a $(q \times q)$ information matrix, where $q = p + p(p+1)/2$ is the number of model parameters. This inversion can be difficult when p is large – for example, if $p = 50$, then $q = 1,325$ and inverting a $(1,325 \times 1,325)$ matrix is challenging, even with modern computing resources.

The EM algorithm does not provide an estimated covariance matrix for $\hat{\theta}$. An appealing approach is to compute the ML estimate of θ on a set of bootstrap samples and estimate the covariance matrix of $\hat{\theta}$ as the sample covariance matrix of the bootstrap estimates. This method gives asymptotically valid standard errors, even though the set of observed variables changes across the units. Computing time is reduced by sorting the cases by pattern before applying EM and grouping cases that have a similar pattern together. I computed bootstrap standard errors for estimates from each simulation replicate in the simulation study reported in Little (1979).

This model also provides an approach to missing data in normal multiple linear regression, with incomplete covariates that are continuous and can be treated as normal. Parameters of the regression are functions of θ. If $g(\theta)$ is one such parameter, then the ML estimate of $g(\theta)$ is simply $g(\hat{\theta})$, where $\hat{\theta}$ is the ML estimate from the EM algorithm given above. The asymptotic standard error can be obtained from the bootstrap distribution of $g(\hat{\theta})$, obtained by computing g on each bootstrap replicate.

An alternative to this ML/bootstrap approach is to apply Bayesian methods, simulating from the posterior distribution of the parameters using methods discussed in the next section.

11.3 Bayesian Monte-Carlo Methods

11.3.1 Introduction

Suppose a model for data Y assumes the data Y have a density $f_Y(Y \mid \theta)$, where $\theta = (\theta_1, ..., \theta_K)$ is a vector with K elements. The objective is a Bayesian analysis, and θ is assigned a prior distribution $\pi(\theta)$. Bayes Theorem implies that the posterior distribution of θ is

$$P(\theta \mid Y) = \pi(\theta) \times f_Y(Y \mid \theta)/c, \tag{11.9}$$

where c is the normalizing constant:

$$c = \int ... \int \pi(\theta) \times f_Y(Y \mid \theta) d\theta_1 ... d\theta_K. \tag{11.10}$$

Bayesian inference about a particular parameter, say θ_1, is based on its marginal posterior distribution:

$$P(\theta_1 \mid Y) = \int ... \int \pi(\theta_1, ..., \theta_K) \times f_Y(Y \mid \theta_1, ..., \theta_K) d\theta_2 ... d\theta_K / c. \tag{11.11}$$

Thus, computing the posterior distribution of θ_1 via (11.10) and (11.11) requires K-dimensional and $(K\text{-}1)$-dimensional integrations. If K is large and the integrals are not analytically tractable, as is the case with many models, then this represents a formidable computational challenge: standard numerical methods for approximating integrals become unwieldy unless K is small, say less than ten or fifteen.

This is where the delightful simplicity of Monte-Carlo methods comes into its own.

Suppose that a computational method can be devised for generating *draws* from the posterior distribution of θ; some approaches to this are discussed in Section 11.3.2. Let $(\theta^{(d)}, d = 1, ..., D)$ be a set of these draws. These draws can be used to approximate the posterior distribution (11.9), although a large value of D may be needed to fill the multidimensional space adequately. However, the draws of θ_1 say $(\theta_1^{(d)}, d = 1, ..., D)$ can also be used to approximate the posterior distribution (11.11), crucially avoiding the need for multidimensional integration. In particular, the posterior mean of θ_1 is approximated by sample mean of the draws:

$$\bar{\theta}_1 = \sum_{d=1}^{D} \theta_1^{(d)} / D,$$

and the posterior variance of θ_1 approximated by the sample variance of the draws:

$$s_1^2 = \sum_{d=1}^{D} \left(\theta_1^{(d)} - \bar{\theta} \right)^2 / (D-1).$$

If the posterior distribution of θ_1 is approximately normal, an approximate 95% posterior credible interval is $(\bar{\theta}_1 \pm 1.96 s_1)$; otherwise, the interval can be approximated using the 2.5th to 97.5th sample percentiles of the draws, or the 95% region with highest posterior density. The normal approximation requires a relatively modest number of draws, say 500; creating the posterior distribution from percentiles requires considerably large values of D, say 5000 or more, but such values of D are often feasible even for large models, given modern computation power and storage.

More generally, suppose Bayesian inferences are required for $g(\theta)$, any scalar function of θ of substantive interest. The posterior distribution of $g(\theta)$ is readily approximated as for θ_1, using the set of draws formed by evaluating the function for each of the drawn values of θ, that is $(g(\theta^{(d)}), d = 1, ..., D)$. This provides a Bayesian analysis for simple or complex functions of θ.

I now discuss two of the many ways of generating draws from a posterior distribution.

11.3.2 Direct Simulation of the Posterior Distribution

For some models, draws from the posterior distribution can be computed directly, that is, without the need for iterative methods. Even when the posterior distribution has a density with known analytic form, simulating draws from it may be advantageous in order to simulate the posterior distribution of functions of the parameters, as just described. I illustrate direct simulation for the Bayesian bootstrap model mentioned in Section 11.1, and for normal data with a monotone missingness pattern, exploiting Anderson's (1957) idea of factoring the likelihood.

Example 11.2. The Bayesian bootstrap

Let $\{x_1, ..., x_n\}$ be a random sample from a variable X, which may be vector valued. Let $d = (d_1, ..., d_K)$ be the set of all possible distinct values of X, and let n_k be the number of x_i equal to d_k, where $\sum_{k=1}^{K} n_k = n$. Then $(n_1, ..., n_K)$ has a multinomial distribution with index n and probabilities

$$\theta = (\theta_1, ..., \theta_K), \text{ where } \theta_k = \Pr(X = d_k \mid \theta), \sum_{k=1}^{K} \theta_k = 1.$$

Assume the following Dirichlet prior distribution for θ:

$$\pi(\theta) \propto \prod_{k=1}^{K} \theta_k^{\ell_k} \left(0 \text{ if } \prod_{k=1}^{K} \theta_k \neq 1 \right). \tag{11.12}$$

The posterior distribution of θ is also Dirichlet, with density

$$p(\theta \mid x_1, ..., x_n) \propto \prod_{k=1}^{K} \theta_k^{d_k + \ell_k} \left(0 \text{ if } x_i \neq d_k \text{ for some } i, k, \text{ or } \prod_{k=1}^{K} \theta_k \neq 1 \right). \tag{11.13}$$

Rubin (1981) describes that a direct draw from the posterior distribution (11.13) can be obtained by the following simple algorithm:

   a. Let $m = n + K + \sum_{k=1}^{K} \ell_k$. Draw $m - 1$ independent uniform random numbers $u_1, ..., u_{m-1} \sim_{\text{ind}} U(0, 1)$.
   b. Let $(g_1, ..., g_m)$ be the gaps generated by the ordered $\{u_i\}$.
   c. Partition the $(g_1, ..., g_m)$ into K collections, the kth having $(n_k + \ell_k + 1)$ elements.
   d. Let P_k be the sum of the g_i in the kth collection, $k = 1, ..., K$.
   e. Then $\theta^{(d)} = (P_1, ..., P_K)$ is a draw from the posterior distribution (11.13).

The posterior distribution is simulated by creating a large number D of draws using this algorithm.

Rubin (1981) shows that this Bayesian analysis, with prior distribution (11.12) with $\ell_k = -1, k = 1, ..., K$, is closely related to the bootstrap method of

Section 11.2, and thus shares the strengths and limitations of that method. In particular, the method assumes a multinomial model that only assigns positive probability to each of the subset of observed values in the data $\{x_1, ..., x_n\}$. This assumption is usually unrealistic, but the method yields useful inferences despite this limitation. Bayesian methods based on Dirichlet process priors relax this assumption and form the heart of modern Bayesian nonparametric methods (e.g., Ghosal & van der Vaart 2017).

Example 11.3. ML and direct simulation of the posterior distribution for normal data with a monotone missingness pattern

Consider a special case of Example 11.1, where the data consist of r complete bivariate observations $\{y_i = (y_{i1}, y_{i2}), i = 1, ..., r\}$ on $Y = (Y_1, Y_2)$, and $n - r$ observations $\{y_{i1}, i = r+1, ..., n\}$ on Y_1 alone. Assuming data are normal, and missingness of Y_2 depends on Y_1 but not Y_2, so the mechanism is missing at random (MAR, Rubin 1976). The loglikelihood is given by

$$\ell_{ign}(\mu, \Sigma) = -\frac{1}{2} r \ln|\Sigma| - \frac{1}{2} \sum_{i=1}^{r} (y_i - \mu) \Sigma^{-1} (y_i - \mu)^T - \frac{1}{2}(n-r) \ln \sigma_{11} - \frac{1}{2} \sum_{i=r+1}^{n} \frac{(y_{i1} - \mu)^2}{\sigma^2},$$

(11.14)

where μ and Σ are the mean and covariance matrix of Y. ML estimates of μ and Σ can be found by maximizing this function with respect to μ and Σ. The likelihood equations based on differentiating (11.14) do not have an obvious solution. However, Anderson (1957) factors the joint distribution of Y_1 and Y_2 into the marginal distribution of Y_1 and the conditional distribution of Y_2 given Y_1. For the ith observation:

$$f\left(y_{i1}, y_{i2} | \mu, \Sigma\right) = f(y_{i1} | \mu_1, \sigma_{11}) f(y_{i2} | y_{i1}, \beta_{20 \cdot 1}, \beta_{21 \cdot 1}, \sigma_{22 \cdot 1}),$$

where $f(y_{i1} | \mu_1, \sigma_{11})$ is the normal distribution with mean μ_1 and variance σ_{11}, and $f(y_{i2} | y_{i1}, \beta_{20 \cdot 1}, \beta_{21 \cdot 1}, \sigma_{22 \cdot 1})$ is normal with mean $\beta_{20 \cdot 1} + \beta_{21 \cdot 1} y_{i1}$ and variance $\sigma_{22 \cdot 1}$. The loglikelihood (11.14) can then be rewritten in the factored form

$$\ell_{ign}(\phi | Y_{obs}) = -\frac{1}{2} r \ln \sigma_{22 \cdot 1} - \frac{1}{2} \sum_{i=1}^{r} (y_{i2} - \beta_{20 \cdot 1} - \beta_{21 \cdot 1} y_{i1})^2 / \sigma_{22 \cdot 1}$$

(11.15)

$$- \frac{1}{2} n \ln \sigma_{11} - \frac{1}{2} \sum_{i=1}^{n} \frac{(y_{i1} - \mu_1)^2}{\sigma_{11}},$$

where $\phi = (\mu_1, \sigma_{11}, \beta_{20 \cdot 1}, \beta_{21 \cdot 1}, \sigma_{22 \cdot 1})^T$ is a one-one function of the original parameters $\theta = (\mu_1, \mu_2, \sigma_{11}, \sigma_{12}, \sigma_{22})^T$ of the joint distribution. If the parameter space for θ is the standard natural parameter space with no prior restrictions, then (μ_1, σ_{11}) and $(\beta_{20\cdot1}, \beta_{21\cdot1}, \sigma_{22\cdot1})$ are distinct, because knowledge of (μ_1, σ_{11}) does not yield any information about $(\beta_{20\cdot1}, \beta_{21\cdot1}, \sigma_{22\cdot1})$. Hence, ML estimates of ϕ can be obtained by independently maximizing the likelihoods corresponding to these parameter subsets. This yields $\hat{\phi} = (\hat{\mu}_1, \hat{\sigma}_{11}, \hat{\beta}_{20\cdot1}, \hat{\beta}_{21\cdot1}, \hat{\sigma}_{22\cdot1})$, where

$$\hat{\mu}_1 = n^{-1} \sum_{i=1}^{n} y_{i1}, \quad \hat{\sigma}_{11} = n^{-1} \sum_{i=1}^{n} (y_{i1}, \hat{\mu}_1)^2, \tag{11.16}$$

the sample mean and sample variance of the n observations $y_{11}, \ldots y_{n1}$, and

$$\hat{\beta}_{21\cdot1} = s_{12}/s_{11}, \ \hat{\beta}_{20\cdot1} = \bar{y}_2 - \hat{\beta}_{21\cdot1}\bar{y}_1, \ \hat{\sigma}_{22\cdot1} = s_{22} - s_{12}^2/s_{11}, \tag{11.17}$$

where $\bar{y}_j = r^{-1}\sum_{i=1}^{r} y_{ij}, s_{jk} = r^{-1}\sum_{i=1}^{r}(y_{ij} - \bar{y}_j)(y_{ik} - \bar{y}_k)$ for $j, k = 1,2$.

The ML estimates of functions of ϕ are simply the functions evaluated at the ML estimates $\hat{\phi}$. For example, the mean of Y_2 is $\mu_2(\phi) = \beta_{20\cdot1} + \beta_{21\cdot1}\mu_1$, so

$$\hat{\mu}_2 = \mu_2(\hat{\phi}) = \hat{\beta}_{20\cdot1} + \hat{\beta}_{21\cdot1}\hat{\mu}_1 = \bar{y}_2 + \hat{\beta}_{21\cdot1}(\hat{\mu}_1 - \bar{x}_1), \ \hat{\beta}_{21\cdot1} = s_{12}/s_{11}. \tag{11.18}$$

ML estimates of the other parameters of the joint distribution are similarly straightforward (Anderson 1957; Little & Rubin 2019, chapter 7).

The Bayesian approach specifies a prior distribution for the parameters and then derives the associated posterior distribution. Specifically, suppose $\mu_1, \sigma_{11}, \beta_{20\cdot1}, \beta_{21\cdot1}$ and $\sigma_{22\cdot1}$ are assumed a priori independent with prior

$$f(\mu_1, \sigma_{11}, \beta_{20\cdot1}, \beta_{21\cdot1}, \sigma_{22\cdot1}) \propto \sigma_{11}^{-a}\sigma_{22\cdot1}^{-c}. \tag{11.19}$$

The choice $a = c = 1$ yields the Jeffreys' prior for the factored density (Box & Tiao 1973).

Applying standard Bayesian theory to the random sample $\{y_{i1}: i = 1, \ldots, n\}$, we have the following results: the posterior distribution of (μ_1, σ_{11}) is such that (1) $n\hat{\sigma}_{11}/\sigma_{11}$ has a chi-squared distribution with $n+2a-3$ degrees of freedom, and (2) the posterior distribution of μ_1 given σ_{11} is normal with mean $\hat{\mu}_1$ and variance σ_{11}/n; applying standard Bayesian regression theory to the random sample $\{(y_{i1}, y_{i2}): i = 1, \ldots, r\}$, the posterior distribution of $(\beta_{20\cdot1}, \beta_{21\cdot1}, \sigma_{22\cdot1})$ is such that (3) $r\hat{\sigma}_{22\cdot1}/\sigma_{22\cdot1}$ has a chi-squared distribution with $r+2c-4$ degrees of freedom, (4) the posterior distribution of $\beta_{21\cdot1}$ given $\sigma_{22\cdot1}$ is normal with mean $\hat{\beta}_{21\cdot1}$ and variance $\sigma_{22\cdot1}/(r\tilde{\sigma}_{11})$, and (5) the posterior distribution of $\beta_{20\cdot1}$ given $\beta_{21\cdot1}$ and $\sigma_{22\cdot1}$ is normal with mean $\bar{y}_2 - \beta_{21\cdot1}\bar{y}_1$ and variance $\sigma_{22\cdot1}/r$; furthermore, (6) (μ_1, σ_{11}) and $(\beta_{20\cdot1}, \beta_{21\cdot1}, \sigma_{22\cdot1})$ are a posteriori independent. For derivations of these results, see, for example, Lindley (1965) or Gelman et al. (2013).

The posterior distribution of any function of the parameters ϕ can be simulated by creating D draws, $d = 1,..., D$, as follows (Little & Rubin 2019, Section 7.3):

1. Draw independently x_{1t}^2 and x_{2t}^2 from chi-squared distributions with $n + 2a - 3$ and $r + 2c - 4$ degrees of freedom, respectively. Draw three independent standard normal deviates $z_{1t}, z_{2t},$ and z_{3t}.

2. Compute $\phi^{(d)} = \left(\sigma_{11}^{(d)}, \mu_1^{(d)}, \sigma_{22\cdot1}^{(d)}, \beta_{20\cdot1}^{(d)}, \beta_{21\cdot1}^{(d)} \right)^T$, where

$$\sigma_{11}^{(d)} = n\hat{\sigma}_{11}/x_{1t}^2, \quad \mu_1^{(d)} = \hat{\mu}_1 + z_{1t}\left(\sigma_{11}^{(d)}/n\right)^{1/2},$$

$$\sigma_{22\cdot1}^{(d)} = r\hat{\sigma}_{22\cdot1}/x_{2t}^2, \quad \beta_{21\cdot1}^{(d)} = \hat{\beta}_{21\cdot1} + z_{2t}\left(\sigma_{22\cdot1}^{(d)}/(r\tilde{\sigma}_{11})\right)^{1/2},$$

$$\beta_{20\cdot1}^{(d)} = \bar{y}_2 - \beta_{21\cdot1}^{(d)}\bar{y}_1 + z_{3t}\left(\sigma_{22\cdot1}^{(d)}/r\right)^{1/2}.$$

3. Compute the corresponding transformation of $\phi^{(d)}$. For example, if the transformation is $\mu_2 = \beta_{20\cdot1} + \beta_{21\cdot1}\mu_1$, then $\mu_2^{(d)} = \beta_{20\cdot1}^{(d)} + \beta_{21\cdot1}^{(d)}\mu_1^{(d)}$.

This direct simulation approach extends to the more general model of Example 11.1, where $(Y_1, Y_2,..., Y_K)$ have a K-variate normal distribution with mean $\mu = (\mu_1,..., \mu_K)$ and covariance matrix $\Sigma = (\sigma_{jk})$, and the pattern of missing data is monotone, that is Y_j observed for all units where Y_{j+1} is observed, for $j = 1,..., K-1$. Each Y_j here can be a set of variables. Also, if $Y_{(1)}$ and $Y_{(2)}$ are any distinct subsets of the variables that make up Y, the draws from the posterior distribution of θ can be transformed to simulate the posterior distribution of the parameters of a multivariate regression of $Y_{(1)}$ and $Y_{(2)}$.

11.3.3 The Gibbs' Sampler

The Gibbs sampler is designed to generate draws from the distribution $p(U)$ of a set of variables $U = (U_1,..., U_p)$, where each U_j can be a set of variables. Suppose that directly drawing from $p(U)$ is not feasible, but it is feasible to draw from the set of conditional distributions $p(U_j | U_{(-j)})$, where U_j is a subset of U and $U_{(-j)}$ denotes the set of U-variables with U_j excluded. Under quite broad conditions, Geman and Geman (1984) show that iteratively updating the draws from the sequence of conditional distributions converges to a draw from the joint distribution. More specifically, let $(U_1^{(0)},..., U_p^{(0)})$ denote initial draws of U. Then at iteration t, draw

$$U_1^{(t)} \sim P\left(U_1 | U_2^{(t-1)}, U_3^{(t-1)},..., U_p^{(t-1)}\right)$$

$$U_2^{(t)} \sim P\left(U_2 | U_1^{(t)}, U_3^{(t-1)},..., U_p^{(t-1)}\right)$$

...

$$U_p^{(t)} \sim P\left(U_p | U_1^{(t)}, U_2^{(t)},..., U_{p-1}^{(t)}\right),$$

where for each draw the most recent draws of the other U values are substituted. Under fairly general conditions, this sequence converges to a draw from the joint distribution as $t \to \infty$.

In Bayesian applications, the components of U are often sets of parameters, and the Gibbs' algorithm is used to simulate the posterior distribution given data D. Gelfand and Smith (1988) discuss how the algorithm can be applied to estimate posterior distributions of multilevel models, as in the following example.

Example 11.4. The Gibbs' sampler for a normal random-effects model

Gelfand & Smith (1988) provide the following example of a model that has proved difficult to handle using empirical Bayes approaches (e.g., Morris 1983b, 1987). Suppose that for $i = 1,...,I$, $j = 1,...,J_i$, and assuming conditional independence of the components:

$$\left(y_{ij} \mid \theta_i, \sigma_i^2\right) \sim_{\text{ind}} N(\theta_i, \sigma_i^2),$$

$$\left(\theta_i \mid \mu, \tau^2\right) \sim_{\text{ind}} N(\mu, \tau^2), \sigma_i^2 \sim IG(a_1, b_1)$$

$$\left(\mu \mid \mu_0, \tau_0^2\right) \sim N(\mu_0, \sigma_0^2), \tau^2 \sim IG(a_2, b_2),$$

where N denotes a normal distribution and IG an inverse gamma distribution, and $(\mu_0, \sigma_0^2, a_1, b_1, a_2, b_2)$ are assumed known, often chosen to reflect diffuse prior information. The data can be reduced to sufficient statistics $\bar{y}_i = \sum_{j=1}^{n_i} y_{ij} / J_i, s_i^2 = \sum_{j=1}^{n_i} (y_{ij} - \bar{y}_i)^2 / J_i$.

The joint distribution of $(Y, \theta, \mu, \sigma^2, \tau^2)$ factors as

$$\left[Y, \theta, \mu, \sigma^2, \tau^2\right] = \left[Y \mid \theta, \sigma^2\right] * \left[\theta \mid \mu, \tau^2\right] * \left[\sigma^2\right] * \left[\mu\right] * \left[\tau^2\right],$$

where

$$\left[Y \mid \theta, \sigma^2\right] * \left[\theta \mid \mu, \tau^2\right] * \left[\sigma^2\right] = \prod_{i=1}^{I} \left[y_i \mid \theta_i, \sigma_i^2\right] * \left[s_i^2 \mid \sigma_i^2\right] * \left[\theta_i \mid \mu, \tau^2\right] * \left[\sigma_i^2\right].$$

Gelfand and Smith (1988) derive the conditional distributions for the Gibbs' sampler as

$$\left[\theta \mid Y, \sigma^2, \mu, \tau^2\right] = N\left(\theta^*, D^*\right), \text{ where } \theta_i^* = \frac{J_i \bar{y}_i \tau^2 + \mu \sigma_i^2}{J_i \tau^2 + \sigma_i^2},$$

$$D_{ii}^* = \frac{\sigma_i^2 \tau^2}{J_i \tau^2 + \sigma_i^2}, D_{ij}^* = 0, i \neq j,$$

$$\left[\sigma_1^2, ..., \sigma_I^2 \mid Y, \theta, \mu, \tau^2\right] = \prod_{i=1}^{I} \left[\sigma_i^2 \mid \bar{y}_i, s_i^2, \theta_i\right], \left[\sigma_i^2 \mid \bar{y}_i, s_i^2, \theta_i\right]$$

$$= IG\left(a_1 + J_i / 2, b_1 + \sum_{j=1}^{J_i} (y_{ij} - \theta_i)^2 / 2\right),$$

$$\left[\mu\mid Y,\theta,\sigma^2,\tau^2\right]=\left[\mu\mid\theta,\tau^2\right]=N\left(\frac{\tau^2\mu_0+\sigma_0^2\sum_{i=1}^{I}\theta_i}{\tau^2+I\sigma_0^2},\frac{\tau^2\sigma_0^2}{\tau^2+I\sigma_0^2}\right).$$

The key point is that all these conditional distributions are either normal or inverse Gamma, for which draws are easily generated.

Example 11.5. The Gibbs' sampler for multivariate normal data with a general pattern of missing data

I now describe a Bayesian analysis of the multivariate normal model in Example 11.3, following the description in Little and Rubin (2019, chapter 11). To simplify the description, assume the conventional Jeffreys' prior distribution for the mean and covariance matrix:

$$p(\mu,\Sigma)\propto|\Sigma|^{-(K+1)/2},$$

The data augmentation (DA) algorithm in Tanner and Wong (1985) applies the Gibbs sampler to the model parameters and the missing data to generate draws from the posterior distribution of $\theta=(\mu,\Sigma)$:

$$p(\mu,\Sigma|Y_{(0)})\propto|\Sigma|^{-(K+1)/2}\exp\left(\ell\left(\mu,\Sigma|Y_{(0)}\right)\right),$$

where $\ell(\mu,\Sigma\mid Y_{(0)})$ is the loglikelihood in Eq. (11.1). Let $\theta^{(t)}=(\mu^{(t)},\Sigma^{(t)})$ and $Y^{(t)}=(Y_{(0)},Y_{(1)}^{(t)})$ denote current draws of the parameters and filled-in data matrix at iteration t. The I (imputation) step of the DA algorithm draws

$$Y_{(1)}^{(t+1)}\sim p\left(Y_{(1)}\mid Y_{(0)},\theta^{(t)}\right).$$

Because the rows of the data matrix Y are conditionally independent given θ, this is equivalent to drawing

$$y_{(1),i}^{(t+1)}\sim p\left(y_{(1),i}\mid y_{(0),i},\theta^{(t)}\right),\tag{11.20}$$

independently for $i=1,\ldots,n$. As noted in the discussion of EM, this distribution is multivariate normal with mean given by the linear regression of $y_{(1),i}$ on $y_{(0),i}$, evaluated at current draws $\theta^{(t)}$ of the parameters. The regression parameters and residual covariance matrix of this normal distribution are readily obtained using the sweep operator (Little & Rubin, Section 7.4.3); for more details see Little and Rubin (2019, Section 11.2.3).

The P (posterior) step of DA draws

$$\theta^{(t+1)}\sim p\left(\theta\mid Y^{(t+1)}\right),$$

where $Y^{(t+1)} = \left(Y_{(0)}, Y^{(t+1)}_{(1)}\right)$ is the imputed data from the I step (11.20). The draw of $\theta^{(t+1)}$ can be accomplished in two steps:

$$\left(\Sigma^{(t+1)}/(n-1) \,|\, Y^{(t+1)}\right) \sim \text{Inv-Wishart}\left(S^{(t+1)}, n-1\right)$$
$$\left(\mu^{(t+1)} \,\big|\, \Sigma^{(t+1)}, Y^{(t+1)}\right) \sim N_K\left(\bar{y}^{(t+1)}, \Sigma^{(t+1)}/n\right),$$

(11.21)

where $(\bar{y}^{(t+1)}, S^{(t+1)})$ is the sample mean and covariance matrix of Y from the imputed data $Y^{(t+1)}$. The posterior distribution of θ can be simulated by iterating between Eqs. (11.20) and (11.21), after a suitable burn-in period to achieve stationary draws.

A by-product of the Gibbs' sampler applied to missing-data problems like this is that the draws (11.20) from the predictive distribution of missing values can be used for another important tool for handling missing data, namely multiple imputation (MI, Rubin 1987). Multiple data sets are generated with missing values replaced by different sets of draws from their predictive distribution. The statistical analysis of interest is applied to each data set, and results are combined using simple MI combining rules, which are simulation-based approximations of Bayesian posterior means and variances.

Chained-equation MI (Raghunathan et al. 2001; van Buuren & Groothuis-Oudshoorn 2011) creates MIs using tailored models for the conditional distribution of each incomplete variable, given the other variables in the data set. While not strictly Bayesian, these methods follow the logic of the Gibbs' sampler by drawing from models for the set of conditional distributions. They provide a level of modeling flexibility that is not attainable by MI based on models for the joint distribution of the incomplete variables. See, for example, Little and Rubin (2019, Section 10.2.4).

11.4 Conclusion

The bootstrap method in Section 11.2 is widely used because of its simplicity and versatility. A variety of elaborations of the method have been developed to improve on the small-sample properties of the basic method, including the accelerated bias-corrected (BC$_a$) method, the approximate bootstrap confidence interval (ABC) method, and the bootstrap t method. See, for example, DiCiccio and Efron (1996), and Davison and Hinkley (1997). While important advances, these methods lose some of the simplicity of the basic method, which is for me a large part of its appeal.

The idea of generating draws from the posterior distribution in a Bayesian analysis, using direct or iterative simulation methods, is key to making the

Bayesian method broadly applicable to multiparameter models. Methods for drawing from posterior distributions have continued to evolve since the development of the Gibbs' sampler discussed in Section 11.3, making Bayesian inference feasible for increasingly complex models; see, for example, Turkman, Paulino and Müller (2019). With the removal of computational obstacles, Bayesian inference is now an important practical alternative to frequentist methods in many areas of statistical application, and I expect it to play an increasingly important role in future advances of the field.

11.5 Some Thought Questions on This Chapter

1. Verify the statement after Eq. (11.1) that if $\hat{\theta}$ has a bias of order $1/n$, then $\hat{\theta}^*_{\text{jack}}$ has a bias of order $1/n^2$.

2. Review the strengths and weaknesses of the bootstrap compared with the jackknife.

3. The bootstrap is commonly regarded as relatively "assumption-free." Rubin (1981) disagrees. Review Rubin's arguments that the bootstrap, and its Bayesian counterpart the Bayesian bootstrap, involve assumptions and can perform poorly when these assumptions are not met.

4. Unlike the EM algorithm for ML estimation with missing data, the Gibbs' Sampler lacks the property of increasing the likelihood at each iteration. Review methods for monitoring convergence of the Gibbs' chain, and the arguments for running multiple sequences rather than a just single chain. (See, e.g., Gelman & Rubin 1992.)

5. The Markov Chain computational methods for Bayesian inference in Section 11.3 are not controversial per se but consider how they lend fuel to the controversies concerning frequentist or Bayesian inference, which play a large role in this book.

6. Explore some of the more recent advances in Bayesian computation in Turkman, Paulino and Müller (2019).

12

Exploratory Data Analysis and Data Science

The papers:

Tukey, J. W. (1962). The future of data analysis. *Ann. Math. Statist.*, 33, 1, 1–67.
Breiman, L. (2001). Statistical modeling: two cultures. *Statist. Sci.* 16, 3, 199–231.

Other reading:

Mitra, N. (2021). Introduction to special issue: Commentaries on Breiman's Two
    Cultures paper. *Observational Studies*, 7, 1, 1–2, and the other papers in that
    volume.
Donoho, D. (2017). 50 years of data science. *J. Comp. Graphical Statist.*, 26, 4, 745–766.
van der Laan, M. J., Polley, E. C. & Hubbard, A. E. (2007). Super learner. *Statist. Applic.*
    Genetics Mol. Biol., 6, 1, Article 25.

12.1 Introduction

Donoho (2017) begins his paper on data science with:

> More than 50 years ago, John Tukey called for a reformation of academic
> statistics. In "The Future of Data Analysis," he pointed to the existence
> of an as-yet unrecognized *science*, whose subject of interest was learning
> from data, or "data analysis."

In the empirical Bayes discussion in Chapter 7, I discussed statistics from
the dual perspectives of mathematics and statistical modeling. In the seminal paper of Tukey (1962), the dual topics are statistics as mathematics or data
analysis. Tukey made his fair share of contributions to mathematical statistics,
one being the studentized range test discussed in Chapter 9, but his uniquely
original contributions were in the realm of data analysis, as in the field of
exploratory data analysis (EDA), which he largely invented (Tukey 1977). In
particular, he is known for simple but ingenious methods for displaying data
like the stem and leaf plot and the box plot, which are ubiquitous in data analysis to this day.

If we add computing to mathematical statistics and data analysis, the
resulting triumvirate creates what I would call "data science," a label much
in vogue at the time of writing, so much so that computer scientists and

DOI: 10.1201/9781003395164-14

statisticians vie for ownership of the term. Tukey was active at the time when computing was taking off, and, working in the computing hotbed of Bell Laboratories, he was well versed in that subject. Donoho and other statisticians view Tukey as the first "data scientist," and the first part of Tukey (1962), summarized here in Section 12.2, provides ammunition for that claim. I'll focus on that part, because the rest of his lengthy paper is more concerned with statistics topics of interest to Tukey at that time, including "spotty data" that lay outside the reach of normal models, multi-response data, factor analysis, and stochastic processes.

Section 12.3 considers another perspective on data science. In his "two cultures" paper, Breiman (2001) argued that algorithmic approaches to data analysis are often superior to formal parametric modeling. The paper was revolutionary, iconoclastic, and controversial, and helped to launch the discipline of machine learning, distinguishing it from the model-based statistical thinking that plays a central role in this book. The paper includes a lively and informative discussion, some of which I include here.

In other reading, Donoho (2017) provides a thoughtful perspective on the field of data science, 50 years after Tukey's remarks on the topic. He writes in the abstract:

> This article reviews some ingredients of the current "data science moment," including recent commentary about data science in the popular media, and about how/whether data science is really different from statistics. The now-contemplated field of data science amounts to a superset of the fields of statistics and machine learning, which adds some technology for "scaling up" to "big data". … Drawing on work by Tukey, Cleveland, Chambers, and Breiman, I present a vision of data science based on the activities of people who are "learning from data," and I describe an academic field dedicated to improving that activity in an evidence-based manner.

Also in other reading, some idea of the strong reactions evoked by Breiman's (2001) paper is seen in the special issue of *Observational Studies* that appeared 20 years after Breiman's paper was published (Mitra 2021 and succeeding papers). That volume includes no less than 30 diverse and fascinating perspectives on Breiman's paper, a selection of which will be discussed in Section 12.4. I give my own take on the topic in Section 12.5.

One of Breiman's (2001) key ideas is that mixing a set of diverse models can lead to better answers than any single model. Van der Laan, Polley and Hubbard (2007) formalize that idea in "superlearner," which uses cross-validation to create an optimal learner as a weighted combination of many candidate learners. The paper shows the adaptivity of this so-called super learner to various true data generating distributions. It is a good example of data science as an amalgam of the fields of statistics and computer science.

12.2 Tukey and Data (Analysis as) Science

Tukey describes three essential characteristics of what makes a science:

(a1) intellectual content,

(a2) organization into an understandable form,

(a3) reliance upon the test of experience as the ultimate standard of validity.

He argues that mathematics does not meet (a3), because its ultimate standard of validity is "an agreed-upon sort of logical consistency and provability." As he sees it, data analysis passes all three tests, and he suggests that data analysis must then take on the characteristics of a science rather than those of mathematics, specifically he states that data analysis must:

> seek for scope and usefulness rather than security; be willing to err moderately often in order that inadequate evidence shall more often *suggest* the right answer; and use mathematical argument and mathematical results as bases for judgement rather than as bases for proof or stamps of validity.

Tukey values mathematical statistics to the extent that it supports data analysis:

> To the extent that pieces of *mathematical statistics* fail to contribute, or are not intended to contribute, even by a long and tortuous chain, to the practice of data analysis, they must be judged as pieces of *pure* mathematics, and criticized according to its purest standards. Individual parts of mathematical statistics must look for their justification toward either data analysis or pure mathematics. Work which obeys neither master, and there are those who deny the rule of both for their own work, cannot fail to be transient, to be doomed to sink out of sight.

In Tukey's (1962) view, Departments of Statistics in the United States tended to overemphasize statistics as mathematics, with data analysis a poor stepsister. He writes:

> Data analysts, even if professional statisticians, will have had far less exposure to professional data analysts during their training. Three reasons for this hold today and can at best be altered slowly: "(c1) Statistics tends to be taught as part of mathematics. (c2) In learning statistics *per se* there has been limited attention to data analysis. (c3) The number of years of intimate and vigorous contact with professionals is far less for statistics Ph.D.'s than for physics (or mathematics) Ph.D.'s."

One aspect of mathematics that Tukey finds at odds with good data analysis is the search for optimal solutions to a problem. He concludes the first section of the paper with:

> The most important maxim for data analysis to heed, and one which many statisticians seem to have shunned, is this: "Far better an approximate answer to the *right* question, which is often vague, than an *exact* answer to the wrong question, which can always be made precise. Data analysis must progress by approximate answers, at best, since its knowledge of what the problem really is will at best be approximate."

12.3 Breiman, Big Data, and Machine Learning

The abstract of Breiman (2001) states his thesis succinctly:

> There are two cultures in the use of statistical modeling to conclusions from data. One assumes that the data are generated by a given stochastic data model. The other uses algorithmic models and treats the data mechanism as unknown. The statistical community has been committed to the almost exclusive use of data models. This commitment has led to irrelevant theory, questionable conclusions, and has kept statisticians from working on a large range of interesting current problems. Algorithmic modeling, both in theory and practice, has developed rapidly in fields outside statistics. It can be used both on large complex data sets and as a more accurate and informative alternative to data modeling on smaller data sets. If our goal as a field is to use data to solve problems, then we need to move away from exclusive dependence on data models and adopt a more diverse set of tools.

Breiman contrasts the data modeling and algorithmic approaches in the context of the problem of assessing the relationship between an outcome Y and a set of predictors $(X_1, ..., X_p)$. Classical parametric models, like the normal multiple regression model, the logistic regression model, and the Cox proportional-hazards model, specify a parametric model:

$$y = f(x_1, ..., x_p \mid \theta) + \varepsilon,$$

where ε is an error term. The focus is on finding a function f that fits the data well, interpreting the parameters in the chosen f, and statistical inference about the parameters, that is, hypothesis tests about components of θ with the nominal size and confidence intervals with the nominal coverage.

In algorithmic inference, the function f is replaced by a "black box," and not the primary interest:

$$y = \boxed{\text{black box}} + \varepsilon.$$

The focus is on prediction accuracy of Y, according to criteria such as mean squared error. Breiman invokes methods such as trees, random forests, and neural nets as examples of this algorithmic approach. He describes a number of consulting projects where he found the algorithmic approach to be valuable, namely: predicting next-day ozone levels, using mass spectra to identify halogen-containing compounds, predicting the class of a ship from high altitude radar returns, using sonar returns to predict the class of a submarine, identity of hand-sent Morse Code, toxicity of chemicals, on-line prediction of the cause of a freeway traffic breakdown, speech recognition, and sources of delay in criminal trials in state court systems. In contrast, Breiman criticizes the perceived infallibility of models in the classical modeling approach: after all, the model is the statistician's model, not nature's model.

In his critique of classical modeling, Breiman invokes two principles, Occam's razor and the Rashomon principle. Occam's razor favors models that are simple, where parameters have a simple interpretation, over complex models with many parameters. While simplicity and interpretability are appealing, reality is often more complicated, and simple models are inferior to more complex models in terms of prediction of Y. For example, single tree models are interpretable but not necessarily good as predictors, whereas random forests, which mix large numbers of distinct tree models, are less interpretable but often do better in terms of prediction. "All models are false but some are useful," to paraphrase Box's famous comment, and it does appear a quite general phenomenon that predictions from mixing a number of disparate models seem to do better than predictions from any one single model. An elegant implementation of this idea is the super-learner proposed by van der Laan et al. (2007).

Rashomon is a famous movie by the great Japanese director Akira Kurosawa, where the same event is interpreted differently from varied perspectives. Breiman's "Rashomon principle" refers to the fact that often statistical models with very different interpretations fit the data nearly equally well. Thus, one model might be picked when another, different model fits the data just as well and is closer to nature's truth. If that is the case, then seizing on one particular model as the truth is promising more than can be delivered.

12.4 Diverse Perspectives on Breiman's Paper

Breiman (2001) includes an informative and lively discussion, and I confine my description to the contributions of two thought leaders, David Cox and

Brad Efron. Cox praises Breiman's "clear statement of the broad approach underlying some of his influential and widely admired contributions," and "striking applications and developments," but he also gently pokes holes at Breiman's critique of "mainstream statistical thinking," which he viewed as somewhat of a caricature. He writes:

> even if we ignore design aspects and start with data, key points concern the precise meaning of the data, the possible biases arising from the method of ascertainment, the possible presence of major distorting measurement errors and the nature of processes underlying missing and incomplete data and data that evolve in time in a way involving complex interdependencies. For some of these, at least, it is hard to see how to proceed without some notion of probabilistic modeling.

Concerning Breiman's focus on prediction, Cox acknowledges the potential utility of an empirical "black box" approach if "prediction is localized to situations directly similar to those applying to the data," citing as examples short-term economic and flood forecasting. However, he argues that some understanding of the underlying process is needed when, as is commonly the case, the conditions for prediction are somewhat different from those from the data, citing as examples "what is the likely progress of the incidence of the epidemic of v-CJD in the United Kingdom, and "the effect on annual incidence of cancer in the United States of reducing by 10% the medical use of X-rays, etc.?" A statistical model that incorporates understanding of the underlying processes may be more successful than a "black box" model in such settings.

Cox goes on to suggest that the real test of modeling approach lies in applications in subject-matter journals, rather than in journals focused on presenting novel statistical methodologies. He writes:

> Thus the real procedures of statistical analysis can be judged only by looking in detail at specific cases, and access to these is not always easy. Failure to discuss enough the principles involved is a major criticism of the current state of theory.

Cox's description of statistical modeling reflects his broad experience of applications:

> Formal models are useful and often almost, if not quite, essential for incisive thinking. Descriptively appealing and transparent methods with a firm model base are the ideal. Notions of significance tests, confidence intervals, posterior intervals and all the formal apparatus of inference are valuable tools to be used as guides, but not in a mechanical way; they indicate the uncertainty that would apply under somewhat idealized, maybe very idealized, conditions and as such are often lower bounds to real uncertainty. Analyses and model development are

at least partly exploratory. Automatic methods of model selection (and of variable selection in regression-like problems) are to be shunned or, if use is absolutely unavoidable, are to be examined carefully for their effect on the final conclusions. Unfocused tests of model adequacy are rarely helpful.

Efron argues that the prediction culture is more prevalent than Breiman suggests. Estimation and testing are a form of prediction, in that, for example, estimating that a drug is superior implies a prediction of how drugs will perform on future patients. Prediction is "only occasionally sufficient"; for example, most surveys have "the identification of causal factors as their ultimate goal." Efron also notes the need for pruning of overgrown regression trees, paralleling the use of regularization via biased estimation in classical regression models.

Twenty years after Breiman's paper, the special issue of *Observational Studies* (Mitra 2021) provides a wealth of retrospective assessments, and space only allows me to comment on a few. Banks (2021) acknowledges the success of Breiman's revolution, given the rise of machine learning in industry that followed Breiman's paper. However, he thinks that careful analysis can yield interpretable models with predictions as good as those from machine learning, citing Rudin and Rudin (2019).

Do algorithmic approaches have underlying models? One view is that the methods associated with machine learning have implied models and make assumptions, and the methods are thus part of a broader concept of modeling rather than a radical break from more traditional parametric models. For example, Bickel (2021) writes:

> We, typically, abstract, using what we think we know and simplifying assumptions, to obtain possible approximations to the processes which we are interested in and which lead to the data. As statisticians we think of the data being generated by some probability mechanism. Think of this as a story. A model is a class of stories we believe the truth belongs to or is close enough to for our purposes. Any member story in the model is characterized by unknowns of different kinds we call parameters. A method is a way we devise of using the data to get closer to the truth, in a way we care about, by estimating some of these parameters... This method is implemented through an algorithm. There are many possible algorithms for any given method. The method can be related to the model structure as, for example, maximum likelihood or Bayes. Or the method and algorithm can come from other considerations, as in classification using neural nets or random forests. For any given model there are many methods, and a method can be applicable to many models. I believe that what Leo calls an algorithmic model is in fact an algorithm implementing a method. Underlying the algorithms he discusses is I claim, a probability model in Leo's and most current examples in machine learning.

Thus, it can be argued than many modern algorithmic methods have underlying statistical models. An early method for machine learning is the lasso (Tibshirani 1996); as discussed in Chapter 8, this regularization method can be viewed as arising from an underlying Bayesian model with a prior distribution, with a tuning parameter that might be estimated by Bayesian methods or the more frequentist approach of cross-validation. In tree-based prediction methods, variables are discretized, and trees created that form groups that are relatively homogenous with respect to the outcome of interest, using an approach analogous to forward variable selection in regression that allows for the inclusion of interactions.

The idea of trees originates in biology (Belson 1959), and a computer program for generating trees was developed under the term automatic interaction detection (A.I.D., see Morgan & Sondquist 1963). A.I.D. achieved a following in the social sciences, and the classification into homogenous groups has appeal for determining alternative treatment protocols in medicine. A.I.D. was criticized by statisticians because of the dangers of overfitting, George Barnard (1974) writing:

> … nowadays with more and more apparently sophisticated computer programs for social science, failure to take account of possible sampling fluctuations is leading to a glut of unsound analyses … I have in mind procedures such as A.I.D., the automatic interaction detector, which guarantees to get significance out of any data whatsoever. Methods of this kind require validation …

Later work by Breiman and others addressed this issue by using cross-validation to prune the trees and avoid overfitting. However, the categorizing of continuous predictors may be inferior to flexible splines, and hierarchical models that give higher priority to main effects and low order interactions have substantial appeal. Mixing trees, as in random forests, can be seen as a more complex model that mixes predictions from the component models; Bayesian approaches provide a principled way of combining models so that better fitting components receive more weight; a good example is Bayesian additive random regression trees (BART, Chipman, George & McCulloch 2010).

Gelman (2021) acknowledges the value of Breiman's methods, but like Efron emphasizes the importance of regularization in complex multiparameter settings, writing:

> A common feature of modern big-data approaches to statistics, including lasso, hierarchical Bayes, deep learning, and Breiman's own trees and forests, is regularization: estimating lots of parameters (or, equivalently, forming a complicated nonparametric prediction function) using some statistical tools to control overfitting, whether by the use of priors, penalty functions, cross-validation, or some mixture of these ideas. …. I take the fact that all these methods can, and do, solve real problems every

day in settings where simple least squares and maximum likelihood would fail, as evidence of the benefit of regularization procedures that enable the fitting of complex response surfaces without immediate fear of overfitting.

Gelman favors the form of regularization offered by Bayesian hierarchical models – see Chapter 7 in this book. He views Breiman's disavowal of Bayesian methods as a blind spot, arguing that Bayesian models can provide a general and practical approach to prediction (via the posterior predictive distribution) as well as inference about model parameters.

In contrast, Baiocchi and Rodu (2021) argue radically that machine learning methods are a "direct rejection of reasoning derived from models." They propose that machine learning provides an alternative which they term "outcome reasoning," the principal vehicle of which is the "Common Task Framework (CTF)", which they describe as follows:

Many readers are likely familiar with the CTF even if the name is unfamiliar; the NetFlix Prize (Bennett et al. 2007) and Kaggle competitions are excellent examples of this framework. Following Donoho (2017), we identify the key features of a problem that exists in the CTF:

- curated data that have been placed in a repository;
- static data (all analysts have access to the same data);
- a well-defined task (e.g., predict y given a vector of inputs x for previously unobserved units of observation);
- consensus on the evaluation metric (e.g., the mean squared error of the predictions from the algorithm on a set of observations); and
- an evaluation data set with observations which is not accessible to the analysts.

Models, and underlying assumptions, are thus replaced by the machine learning algorithm, and success determined by whether the algorithm does well on the defined task. This perspective avoids the need for understanding the process under analysis, and the term "model-free" appears in some of the comments in the *Observational Studies* volume. But one might argue, as in Cox's (2001) discussion of Breiman, that often statistics has more general aims than success at such a limited task. Vansteelandt (2021) writes:

I agree with other discussants of Breiman (2001) that prediction problems constitute only a minority of the problems that scientists face. A majority of empirical studies rather aims to develop insight into the effects of exposures on certain outcomes, into causal mechanism, or at a less ambitious level, into associations (e.g., detecting subgroups of the population that are more vulnerable to certain diseases).

Observational Studies is a journal devoted to causal inference, and a number of commenters view Breiman's paper through that lens. For example, Cruz-Cortés et al. (2021) regard the arrow in Breiman's picture from predictors x to outcome y and state that "the directionality of the arrows and the structure of association itself already require a level of modeling assumptions." It seems hard to introduce causality without a causal model, such as the Neyman/Rubin causal model discussed in Chapter 14. The fact that x is being used to predict y does have the implication that x predates y; on the other hand, this does not mean that x causes y, and I am not sure that pure prediction necessarily implies any assumption of causality.

Rudin (2019) argues for "interpretable" machine learning models over "black box" models, writing that

> the lack of transparency and accountability of predictive models can have (and has already had) severe consequences; there have been cases of people incorrectly denied parole, poor bail decisions leading to the release of dangerous criminals, [machine learning]-based pollution models stating that highly polluted air was safe to breathe, and generally poor use of limited valuable resources in criminal justice, medicine, energy reliability, finance and in other domains.

I make some comments on interpretability in the next section.

12.5 Some Personal Perspectives

Breaking the mold of classical models. Breiman is to be commended for challenging the statistics profession to think more broadly than classical parametric models.

The prediction perspective. I like Breiman's prediction perspective, which focuses on real outcomes rather than parameters that only live within a simplified model. I like to say that statistics is basically about "predicting the stuff you don't observe, with appropriate measures of uncertainty."

I like models. Rather than making sharp divisions, I tend to view these camps as different styles of statistical modeling, with strengths and weaknesses depending on whether the setting is purely prediction or requires interpretability. I do not view the machine learning approach as "model-free," because in nearly all applications of statistics, there is a target population that lies beyond the specific data under analysis, and a model is needed to make inferences about that population, with some assessment of uncertainty. I also like Bayes as an overarching paradigm for statistical inference, as discussed in Chapter 6.

The dual roles of statistics. Statistical analysis generally has dual roles – discovering relationships and inference about parameters and prediction of outcomes. For the former, simple parametric models that capture the main

features of a problem may be more useful than complex models whose interpretation is not clear. For prediction, on the other hand, complex models such
as neural nets or the super-learner perform well, at least for situations that
are similar to those under which the data are collected.

Rashomon and the role of design. Breiman (2001) wrote that statistics starts
with data, but I'd say that statistics starts with a problem, and a design for data
collection to address it. The importance of design is illustrated by Breiman's
Rashomon principle, where models with different interpretations having similar fits. The issue is largely avoided by randomized trials where the assignment of treatments is random, as discussed in Chapter 14. It is much more of
an issue for observational data, where treatment differences can be due to a
variety of potential confounding factors, some perhaps not even measured.

If dissimilar models fit equally well and parameter interpretation is the
goal rather than prediction, the statistical analysis should bring out this
ambiguity, rather than focusing all attention on the best-fitting of the candidate models. As an example, in my applications of regression to World
Fertility Survey data (Little 1988), I preferred another strategy for selecting
predictor variables that is more descriptive in flavor. The effect of a particular
predictor variable on the outcome variable is studied for a variety of choices
of concomitant variables, which are introduced into the model according to a
prespecified ordering. For example, Table 12.1 shows an analysis of regional
differentials in current contraceptive use in Thailand, from Cleland, Little
and Pitaktepsombati (1978). The first row gives the unadjusted regional
means, and subsequent rows give the adjusted means from a sequence of

TABLE 12.1

Percentage of Currently Married, Nonpregnant, Fecund Women in Thailand
Currently Using Contraception by Region, Adjusted for Indicated Controls by
Regression

Step	Controls Added	Region of Residence					
		Bangkok	North	Northeast	South	Central	Mean
0	—	55	53	33	19	54	43
1	No. living children	56	53	32	19	54	43
2	Age	56	54	31	19	54	43
3	Urbanity	44	55	32	19	55	43
4	Husband's education	44	55	32	21	54	43
5	Husband's occupation	43	55	33	22	52	43
6	Standard-of-living index	40	55	35	22	50	43
Regional percentage distribution		6.7	25.8	35.3	9.9	22.3	100.0

Source: Little (1988). Copyright © 2012 American Statistical Association, reprinted by permission of Informa UK Limited, trading as Taylor & Francis Group, www.tandfonline.com
on behalf of 2012 American Statistical Association.
Note: The sample size is 2141.

regressions, where controls are added in a chosen sequence. Similar tables were constructed for the other predictor variables in this list, with the variable of interest always being the first introduced into the regression.

Prediction does not imply causation. Rudin (2019) argues for interpretable models but does interpretable imply causation? We describe a regression coefficient in a multiple linear regression as the effect on the mean of Y of increasing X by one unit, holding other variables fixed; but "effect" suggests causation, as in the Neyman/Rubin causal model of Chapter 14, whereas with observational data, we can only infer association, not causation. In focusing on interpretability as in Rudin (2019), we need to be careful to resist the temptation to infer causality when that is not justified by the design.

Occam's Razor. All models are approximations of the truth, and the complexity of the model should depend on the sample size – larger data sets allow for more complicated models. This is one reason why asymptotic properties, where the model is fixed and the sample size increased to infinity, have always struck me as a bit unrealistic. Fitting a complex model to large data sets may yield excellent predictions but may obscure the clearer interpretation of a simpler model.

As an example, when working at the World Fertility Survey, my boss Sir Maurice Kendall emphasized the misleading effects of fitting an additive regression model to data when interactions among the predictor variables are present. He argued against presenting any additive model where statistically significant interactions can be shown to exist. My own position was less rigorous. I argued (Little 1988) that additive models provide useful summary measures in the presence of interactions, providing the results are not viewed as giving the whole picture. In a very large data set, even very small and unimportant interactions might be highly significant, but an additive model conveys the important findings. What matters is not the P-value but the size of the estimated effects.

12.6 Some Thought Questions on This Chapter

1. Do you find EDA a necessary and useful step in your work? Why or why not?

2. Define the terms "Data Science," "Machine Learning," "Artificial Intelligence."

3. Isn't statistics "data science?" How is the field of statistics changing with the advent of the "big data" revolution?

4. For a data science problem with which you are familiar, consider whether there is an underlying population that is the target for inference.

5. Sample the commentaries on the Breiman paper in the special volume of *Observational Studies* (Mitra 2021). Classify them according to whether they are written by members of the statistics or computer science communities. Which do you find compelling?

6. Why do you think ensembling works better for prediction?

7. How can Bayes Theorem be used for ensembling a set of models? Outline the difference between the Bayesian approach to ensembling and the approach in van der Laan, Polley and Hubbard (2007).

Part III

Topics in Design

13

Randomization in Survey Sampling

The paper:

Neyman, J. (1934). On the two different aspects of the representative method: the method of stratified sampling and the method of purposive selection. *J. Roy. Statist. Soc.*, 97, 4, 558–625.

Other reading:

Little, R. J. (2012). Calibrated Bayes: an alternative inferential paradigm for official statistics (with discussion and rejoinder). *J. Official Statist.*, 28, 3, 309–372.

Little, R. J. (2014). Survey sampling: past controversies, current orthodoxies, and future paradigms. In *Past, Present and Future of Statistical Science*, Lin, X., Banks, D. L., Genest, C., Molenberghs, G., Scott, D. W. and Wang, J.-L. (eds.). CRC Press.

Rubin, D. B. (1976). Inference and missing data. *Biometrika*, 63, 581–592.

13.1 Introduction

In 1976, I joined the largest social science project in the world. The World Fertility Survey (WFS) interviewed some 350,000 women in 42 developing and 20 developed countries, collecting data on human fertility and its correlates (e.g., Lightbourne, Singh and Green 1982). The first director of the WFS was Sir Maurice Kendall, author with Allan Stuart of a classic early treatise on statistics (Stuart 2010), as well as witty poetry (Kendall 1959) and children's books. Kendall promoted the WFS by proclaiming that the surveys in the developing countries, unlike some counterparts in the developed world, were "scientific," by which he meant that they were collected using probability sampling designs. Despite the many challenges of probability sampling in the modern world, it remains a cornerstone of data collection in most government statistical agencies. My time at the WFS (e.g., Little 1988) piqued my interest in survey sampling as a field.

Many survey samplers view Neyman (1934) as a seminal paper in the development of probability sampling. The paper is long and complex, and in Section 13.2, I confine attention to what I see as the key ideas that made it famous, omitting some of the details. An interesting sidelight is that a seminal idea in the entire field of statistics – confidence interval estimation – is introduced in the appendix of the paper. This sparked controversy in its own right – Bowley in the discussion of Neyman (1934) famously refers to confidence limits as a "confidence trick." In Section 13.2, I focus on the central question of why probability sampling is a such an important concept, drawing on the description in Little (2014) in the other reading.

A major controversy in survey sampling concerns two radically different approaches to statistical inference. The prevailing orthodoxy is *design-based* (or randomization) inference, where survey items are not assigned a distribution, and statistical uncertainty derives from the probability distribution that determines sample selection. The alternative approach is model-based inference, where inference is based on a statistical model for the survey items, as is common in other application areas of statistics. The two approaches cause confusion, one instance being the role of sampling weights in regression. The issue relates to the more general question about frequentist versus Bayesian inference in statistics at large. While Neyman (1934) is often considered seminal by design-based statisticians, Neyman never explicitly states that he regards population values as fixed, and his references to Student's *t* distribution suggest to me that he had a model in mind. In Sections 13.3–13.5, I summarize this debate, arguing as in Little (2012, 2022) for a model-based – more specifically calibrated Bayesian – approach that incorporates features of the survey design into the model.

It is often thought that model-based inference denies the importance of probability sampling, because the sampling distribution is not the basis for the inference. However, if the selection indictors are included as part of the model, random sampling has the decisive advantage of making the selection mechanism ignorable, which greatly simplifies the modeling task. The key idea is in the paper by Rubin (1976) in the other reading. That paper concerns missing data, but the ideas are directly applicable to survey sampling, by treating sample selection as creating a form of missing data, with the data on non-selected units being unobserved and therefore, in a sense, missing. Rubin's (1976) formulation is couched in terms of sample selection in Section 13.4. Rubin (1978), the topic of the next chapter, expands these ideas to also address the role of randomization in the allocation of treatments.

The main focus of this chapter is on inference about *finite population quantities*, as opposed to parameters of models. A good feature of this is that finite population quantities are real, whereas parameters such as regression coefficients in a multiple regression are quantities in an idealized statistical model. Survey samplers call inference for model parameters "analytic" survey inference. When the interest is on such parameters, we can focus on a corresponding finite population quantity Q, namely the estimate of the parameter of interest if the model was fitted to data for the whole population, according to some agreed fitting method such as least squares or maximum likelihood (ML). A useful feature of this construction is that Q is then also a real quantity (Little 2004).

13.2 The Main Contributions of Neyman's (1934) Paper

The simplest application of probability sampling is simple random sampling (srs) without replacement. Given a population of size N, a unit is selected at random from the population. The selected unit is removed from the population and the process is repeated until n units have been selected. The process is analogous to random selection of numbered balls from a container, as when winning lottery numbers are being selected. In this method, the probability that any one unit is selected is n/N, and the probability that any sample of size n is selected is $n!(N-n)!/N!$. Samples of size other than n have no chance of being selected.

The distribution of a population characteristic, such as age or sex, in the resulting sample may differ substantially by chance from the distribution in the population. If this distribution is known, for example, from an external data source such as a census, then the sample would seem more representative of the population if it was chosen purposively so that the sample and population distributions of the characteristic match. A common form of this is "quota sampling," where interviewers are sent out to survey people whom they encounter and are asked to meet a certain target count in each cell defined by demographic characteristics such as age and sex. Unlike probability sampling, this is not easy to formalize mathematically, and in practice, many forms of non-probability sampling occur.

Neyman (1934) suggests stratified sampling as a development of srs that retains the ability of purposive sampling to match the sample to the population on known characteristics but still has the objective aspects of probability sampling. The population is stratified on the known characteristic, and a srs is selected within each stratum so that the proportion of the sample and population in each stratum is the same (aside from rounding error, because the sample size in each stratum must be a whole number).

Neyman also allows the sampling fraction within each stratum to vary and proposes what has become known as "Neyman allocation," where the sampling fractions are chosen to be proportional to the estimated variance of a survey outcome in each stratum – strata with higher variance are thus assigned a higher sampling fraction. Units in each stratum are then weighted by the inverse of their selection probability, leading to unbiased estimates. For a given sample size, this yields an estimate of the mean with the smallest variance, if the within-stratum estimates of sampling variance were in fact correct. In practice, prior knowledge of these sampling variances is often limited, but Neyman allocation nevertheless remains an important concept in survey design.

In general, a probability sample is characterized by two properties: every *sample* has a known probability of being selected, and every *unit* in the population has a known and *positive* probability of being selected, that is, every unit has a chance of being included. This definition covers srs, stratified srs, and more elaborate sampling designs. Neyman (1934) helped to spur the development of so-called "complex" sample designs, involving stratification, systematic sampling, cluster sampling, and multistage sampling, with potentially unequal sampling fractions within each stage. These features allow the probability sampling of populations where simple random sampling is too expensive or simply not feasible, because a list of population units is not available. Examples are the American Community Survey and Current Population Survey conducted by the U.S. Bureau of the Census.

13.3 Design-based Survey Inference

For a survey of a finite population with N units, let $S = (S_1, ..., S_N)$ where S_i is the selection indicator for the ith unit, taking value 1 when the ith unit is selected and 0 otherwise. Let $Y = (y_1, ..., y_N)$ where y_i is the set of survey variables and let Z represent design information such as stratum or cluster indicators, and z_i the value of Z for unit i. Consider inference about a finite population quantity $Q(Y, Z)$, for example, the population total $Q(Y, Z) = \sum_{i=1}^{N} y_i$, where $Y = (y_1, ..., y_N)$.

The design-based approach treats Y and Z as fixed and bases inference on the distribution of statistics in repeated sampling from the distribution of S. Specifically, a consistent estimator $\hat{q}(S, Z, Y_{\text{inc}})$ of $Q(Y, Z)$ is defined, which is a function of Z and the sampled values Y_{inc} of Y. A consistent estimate of the variance of $\hat{q}$ in repeated sampling of S, say $\hat{V}(Y_{\text{inc}}, S, Z)$, is also derived. A finite-population central limit theorem implies that $\hat{q}$ is normal in large samples, leading to $100(1 - \alpha)\%$ confidence intervals for Q of the form $(\hat{q} \pm z_{(1-\alpha/2)} \sqrt{\hat{V}})$, where $z_{(1-\alpha/2)}$ is the $100(1 - \alpha/2)$ percentile of the standard normal distribution.

> **Example 13.1. Inference for a population mean**
> **from a simple random sample**
>
> Suppose the target parameter is the population mean $Q(Y) = \bar{Y} = (Y_1 + \cdots + Y_N)/N$, and a simple random sample of size n is selected. A natural estimate of $\bar{Y}$ is the sample mean $\bar{y} = \sum_{i=1}^{N} Y_i S_i / n$, the sample indicators S_i extracting the sampled values of Y. This estimate is design-unbiased, because taking expectation over the distribution of S, $E(\sum_{i=1}^{N} Y_i S_i / n \mid Y) = \sum_{i=1}^{N} Y_i / N$. Routine calculations yield the variance of $\bar{y}$ to be $V = (1 - f)S_y^2 / n$, where $f = n/N$ is the sampling fraction and $S_y^2 = \sum_{i=1}^{N} (Y_i - \bar{Y})^2 / (N - 1)$, the population variance of Y. The estimate of V is then $\hat{V} = (1 - f)s_y^2 / n$, replacing the population variance by the sample

variance $s_{\bar{y}}^2 = \sum_{i=1}^N S_i (Y_i - \bar{y})^2 / (n-1)$. The resulting 95% confidence interval for $\bar{Y}$ is then

$$\bar{y} \pm 1.96\sqrt{(1-f)}\left(s_Y/\sqrt{n}\right), \tag{13.1}$$

where 1.96 is the 97.5th percentile of the normal distribution. Note that the factor $\sqrt{(1-f)}$ reflects the inference for the finite population mean, resulting in the variance tending to zero as $n \to N$.

The interval (13.1), and design-based inference in general, assumes large samples. In small samples and a continuous Y, the normal percentile would often be replaced by the percentile of a t distribution, reflecting uncertainty in estimating the variance. But this in effect assumes a normal model for Y, and so is not strictly design-based inference.

Example 13.2. Design-based inference for the mean from a stratified sample

For a sample where unit i is selected with probability $\pi_i = E(S_i | Y, Z)$ and the population size N known, a standard unbiased estimate of the population mean $\bar{Y}$ is the Horvitz-Thompson (1952) estimate

$$\bar{y}_{HT} = N^{-1}\sum_{i=1}^N S_i Y_i / \pi_i, \tag{13.2}$$

which weights sampled units by the inverse of their selection probabilities. A resulting 95% confidence interval is $\bar{y}_{HT} \pm 1.96\hat{v}$, where $\hat{v}$ is a variance estimate that depends on specifics of the sample design. When N is not known, the usual approach replaces $\bar{y}_{HT}$ by the Hájek (1971) estimator

$$\bar{y}_{HK} = \frac{\displaystyle\sum_{i=1}^N S_i Y_i / \pi_i}{\displaystyle\sum_{i=1}^N S_i / \pi_i}, \tag{13.3}$$

where N is effectively being estimated by $\sum_{i=1}^N S_i / \pi_i$.

For the stratified sampling design suggested by Neyman (1934), with selection fraction $f_j = n_j / N_j$ in stratum j, $\bar{y}_{HT}$ estimate reduces to the stratified mean

$$\bar{y}_{HT} \equiv \bar{y}_{st} = \sum_{j=1}^J P_j \bar{y}_j, \tag{13.4}$$

where $\bar{y}_j$ is the sample mean and $P_j = N_j / N$ is the proportion of the population in stratum j.

The estimated variance of the stratified mean is

$$\hat{\sigma}_{st}^2 = \sum_{j=1}^{J} P_j^2 (1 - f_j) s_j^2 / n_j, \qquad (13.5)$$

where s_j^2 is the sample variance of Y in stratum j. A 95% confidence interval for $\bar{Y}$ is thus $\bar{y}_{st} \pm 1.96 \hat{\sigma}_{st}$. This approach is satisfactory if the sample size in each stratum is large enough to estimate the stratum variances with acceptable precision. Otherwise, a model-based approach is needed to reflect uncertainty in the stratum variance estimates.

Example 13.3. Horvitz-Thompson meets Basu's elephants

The HT estimator is widely used but can perform poorly if the weights are variable. Basu (1971) provides the following extreme but amusing example:

> The circus owner is planning to ship his 50 adult elephants and so he needs a rough estimate of the total weight of the elephants. As weighing an elephant is a cumbersome process, the owner wants to estimate the total weight by weighing just one elephant. Which elephant should he weigh? So the owner looks back on his records and discovers a list of the elephants' weights taken 3 years ago. He finds that 3 years ago Sambo the middle-sized elephant was the average (in weight) elephant in his herd. He checks with the elephant trainer who reassures him (the owner) that Sambo may still be considered to be the average elephant in the herd. Therefore, the owner plans to weigh Sambo and take $50y$ (where y is the present weight of Sambo) as an estimate of the total weight $Y = Y_1 + Y_2 + \cdots + Y_{50}$ of the 50 elephants. But the circus statistician is horrified when he learns of the owner's purposive sampling plan. "How can you get an unbiased estimate of Y this way?" protests the statistician. So, together they work out a compromise sampling plan. With the help of a table of random numbers they devise a plan that allots a selection probability of 99/100 to Sambo and equal selection probabilities of 1/4900 to each of the other 49 elephants. Naturally, Sambo is selected and the owner is happy. "How are you going to estimate Y?" asks the statistician. "Why? The estimate ought to be $50y$ of course," says the owner. "Oh! No! That cannot possibly be right," says the statistician, "I recently read an article in the *Annals of Mathematical Statistics* where it is proved that the Horvitz–Thompson estimator is the unique hyperadmissible estimator in the class of all generalized polynomial unbiased estimators." "What is the Horvitz–Thompson estimate in this case?" asks the owner, duly impressed. "Since the selection probability for Sambo in our plan was 99/100," says the statistician, "the proper estimate of Y is $100y/99$ and not $50y$." "And, how would you have estimated Y," inquires the incredulous owner, "if our sampling plan made us select, say, the big elephant Jumbo?" "According to what I understand of the Horvitz–Thompson estimation method," says the unhappy statistician, "the proper estimate of Y would then have been $4900y$, where y is Jumbo's weight." That is how the statistician lost his circus job (and perhaps became a teacher of statistics!).

Note that the Hájek estimator (13.3) estimates the mean weight by y, and hence the total weight as $50y$, as the circus owner suggested. Why is the HT estimator sensible in Example 13.2 but manifestly silly in Example 13.3? Why is the Hájek estimator more reasonable than HT in that example, even though N is known? For answers, we turn to the modeling approach to surveys, which is the topic of the next section.

13.4 The Modeling Approach to Survey Inference

The key idea of Rubin's (1976) paper on missing data is including variables M, which indicate whether values are observed or missing, as random variables in the model – that is, if the data are Y, the full model concerns not just the distribution of Y but the joint distribution of Y and M. We can apply Rubin's approach to survey sampling by replacing the missingness indicator M by the indicator for sample selection, S. With the notation of the previous section, a parametric model for the joint distribution of survey variables Y and selection indicators S is then

$$f_{S,Y|Z}(S,Y\,|\,Z,\theta,\psi) = f_{Y|Z}(Y\,|\,Z,\theta)f_{S|Y,Z}(S\,|\,Z,Y,\psi), \qquad (13.6)$$

where $f_{Y|Z}$ represents the density of survey variables Y indexed by unknown parameters θ, and $f_{S|Y,Z}$ represents the model for selection indexed by unknown parameters ψ. The model is used to predict the non-sampled values of Y, leading to inferences for the target population quantities Q.

Rubin (1976) calls the selection mechanism *ignorable* if inference can be based on the model for Y alone, that is $f_{Y|Z}(Y\,|\,Z,\theta)$, greatly simplifying the modeling process. Rubin gives sufficient conditions for ignoring the selection mechanism for both frequentist and Bayesian inference. The key condition is that selection depends on Y and Z only through their observed values, (Y_{inc},Z). The conditions are satisfied for a probability sample, where the sampling distribution is known and does not depend on Y, that is,

$$f_{S|Y,Z}(S\,|\,Z,Y,\psi) = f_{S|Z}(S\,|\,Z). \qquad (13.7)$$

For non-probability samples, on the other hand, ignorability is a strong condition, and violations lead to bias and poor inferences. Thus, the importance of probability sampling is to make the selection process ignorable and hence simplify the modeling process and reduce sensitivity to model misspecification.

The focus in the modeling approach is on prediction of the non-sampled values. In frequentist superpopulation modeling (e.g., Valliant, Dorfman & Royall 2000), the parameters in models are treated as fixed; in Bayesian survey modeling, these parameters are assigned a prior distribution, and inference for $Q(Y, Z)$ are based on its posterior predictive distribution given the data. In large samples, the prior distribution plays a minor role, and the two approaches yield similar answers for comparable models; in particular the ML estimate of a parameter is essentially the mode of the posterior distribution under a uniform prior, and as such has a Bayesian interpretation. In small samples, the Bayesian approach reflects uncertainty about the model parameters when they are integrated out of the posterior distribution. This approach to propagating error in parameters allows Bayesian inferences for judiciously chosen models and priors to be better calibrated than inferences from superpopulation modeling inferences, in the sense discussed in Chapter 6. So, I like to say that "superpopulation modeling is super, but Bayes is better."

Example 13.4. Model-based inference for the population mean from a stratified sample (Example 13.2 continued)

The conditions for ignoring the selection mechanism condition on design variables Z, and a key feature of well-calibrated model-based inferences is that the model deals appropriately with this design information. In particular, for inference about the mean from the stratified sampling design in Example 13.2, the model needs to allow for different means across strata. With the notation of Example 13.2, a standard normal model, with weakly informative Jeffreys' prior on the mean and variance in each stratum, is

$$\left(y_{ji} \mid \mu_j, \sigma_j^2\right) \sim_{\text{ind}} G\left(\mu_j, \sigma_j^2\right), p\left(\mu_j, \sigma_j^2\right) \sim_{\text{ind}} 1/\sigma_j^2, \qquad (13.8)$$

where $G(a, b)$ denotes the normal (Gaussian) distribution with mean a and variance b.

Conditional on the stratum variances $\{\sigma_j^2, j = 1, ... J\}$, the posterior mean of the overall mean $\bar{Y}$ is the stratified mean (13.4), with posterior variance given by (13.5). Thus asymptotically, the posterior credible interval is the same as the confidence interval from the design-based approach. Integrating over the posterior distribution of $\{\sigma_j^2, j = 1, ... J\}$ yields inferences that propagate uncertainty from estimating the unknown variances. Specifically, σ_j^2 is a scaled inverse chi-squared distribution, and the posterior distribution of $\bar{Y}$ is a mixture of t distributions. When $J = 2$, the model is the same as that for Fisher's solution to the Behrens-Fisher problem discussed in Chapter 4.

The sample size in each stratum has to be at least 3 for the posterior distribution of the variances to be proper. Various modifications of (13.8) address the situation where the sample sizes in the strata are too small. The variances in each stratum can be assumed equal, yielding a pooled estimate of the variance, or a random-effects model on the variances can smooth the individual stratum variance toward a pooled variance. The point is that the Bayesian approach has various solutions to propagating error in estimating these variances, whereas the design-based approach assumes large samples.

Example 13.5. Model-based inference for samples with unequal probabilities of selection

For designs with unequal selection probabilities, classic papers critiquing the modeling approach to surveys (Hansen, Madow & Tepping 1983; Kish & Frankel 1974) do not include the selection probabilities in the model, yielding inferences that are vulnerable to model misspecification. The selection probabilities play an important role in robust model-based inference, but as model covariates rather than as sampling weights.

Consider, for example, inference about a population mean $\bar{Y}$. If y_i is the value of a survey variable Y and π_i is the selection probability for unit i, the modeling approach incorporates the selection probabilities by regressing y_i on π_i. The strength of relationship between y_i and π_i then moderates how the selection probability affects the estimator – if the relationship is weak, the regression coefficient of π_i is small and the sampling weight has little influence. This results in more efficient estimates.

A linear regression of y_i on π_i is vulnerable to bias if the linearity assumption is violated, but the impact of misspecification can be reduced by choosing a model that results in a design-consistent estimate, in the sense that the model-based estimate converges to the true population quantity as the sample size is increased, whether or not the model is well-specified (e.g., Isaki & Fuller 1982). Many models satisfy this requirement; see, for example, Firth and Bennett (1998).

In probability proportional to size (PPS) sampling, the covariate Z is the size of the unit and $\pi_i = \min(cz_i, 1)$. Now Z is a continuous variable, and weighting and regression may yield different answers. The approaches can be unified by considering models that yield the design-weighted estimates when used to predict the non-sampled units. In particular, ignoring finite population corrections, the HT estimate (13.2) is the posterior mean for the "Horvitz-Thompson model":

$$\left(y_i \mid z_i, \beta, \sigma^2\right) \sim_{\text{ind}} G\left(\beta z_i, \sigma^2 z_i^2\right), \quad p(\beta, \sigma) \propto 1/\sigma, \tag{13.9}$$

and the Hájek estimate (13.3) is the posterior mean for the "Hájek model":

$$\left(y_i \mid z_i, \beta, \sigma^2\right) \sim_{\text{ind}} G\left(\beta, \sigma^2 z_i\right), \quad p(\beta, \sigma) \propto 1/\sigma. \tag{13.10}$$

These underlying models describe situations where the corresponding design-based estimates are optimal. However, they involve strong parametric assumptions. A robust Bayesian modeling approach embeds these models within a larger model, such as the penalized spline model proposed in Zheng and Little (2005):

$$(y_i \mid z_i, \beta, \sigma^2) \sim_{\text{ind}} G\left(\beta_0 + \sum_{j=1}^{p} \beta_j z_i^j + \sum_{\ell=1}^{m} \beta_{\ell+p}\left(z_i - \kappa_\ell\right)_+^p, \sigma^2 z_i^\alpha\right), \tag{13.11}$$

$$(\beta_{l+p} \mid \tau) \underset{\text{iid}}{\sim} N(0, \tau^2), l = 1, ..., m; \quad p(\beta_0, ..., \beta_p, \alpha, \sigma, \tau) \propto 1/\sigma, 0 < \alpha < 2,$$

where the constants $\kappa_1 < ... < \kappa_m$ are selected fixed knots, and $(u)_+^p = u^p$ if $u > 0$ and 0, otherwise (e.g., Ruppert, Wand & Carroll 2003). Chen et al. (2017) provide simulations suggesting that the model (13.11) can yield substantial gains over HT or Hájek estimation, both in terms of efficiency and closer to nominal (frequentist) confidence coverage in moderate samples. The model (13.11) is readily expanded to include other auxiliary variables measured for all the population units, and the flexibility of small-sample inferences increased by including proper prior distributions for the model parameters.

Returning to Basu's elephants in Example 13.3, the HT model (13.9), where the expected weight for elephant i is proportional to the probability of selection, is clearly unreasonable, whereas the Hájek model (13.10) is much more plausible. This is why the Hájek estimator is more sensible than the HT estimator. It is not a question of whether N is known, it is a question of which is the more sensible model.

13.5 Design-based versus Model-based Inference: the Calibrated Bayes Philosophy

The survey sampling literature features many lively arguments (e.g., Brewer, 2013; Kish, 1995; Little, 2014; Smith, 1976, 1994) between "design-based" inference, where inference is based on the sampling distribution (2) and "model-based" inference, where inference is based on model distribution $f_{Y|Z}(Y \mid Z, \theta)$ if selection is ignorable, or on the full model distribution (13.6) if selection is nonignorable.

Good features of the design-based approach are that it avoids explicit dependence on a model for the population values. Models can motivate the choice of estimator, but the inference remains design-based, hence somewhat nonparametric. Design-based properties like design consistency confer robustness, since they apply regardless of the validity of a model. Design weights can be applied uniformly to a set of outcomes, simplifying the computing. On the other hand, true probability samples are harder and harder to come by, noncontact and nonresponse are increasing, and face-to-face interviews are increasingly expensive. The theory of design-based inference is basically asymptotic, providing limited tools for small sample problems like small area estimation. Although not explicitly model-based, models are needed to motivate the choice of estimator. If implicit models are unreasonable, then the resulting inferences can be very poor in moderate samples (Example 13.3 being an extreme case).

The model-based approach is flexible and provides a unified approach for survey problems, including nonresponse and response errors, small area models, and combining information across data sources. The approach moves survey sample inference closer to mainstream statistics, in that disciplines like economics, demography, and public health rely on statistical modeling. Models can be formulated that incorporate sample design features, and the Bayesian approach is not asymptotic, providing superior small-sample inferences. Probability sampling is justified as making sampling mechanism ignorable and improving robustness to model misspecification.

Issues with the model-based approach include more explicit dependence on the choice of model, which has subjective elements; no models are perfect, and thus suffer from a potential lack of robustness to model misspecification. The approach can entail more complex computation, in that models needed for all survey outcomes.

The calibrated Bayes approach discussed in Chapter 6 seeks the best of both worlds – inferences are based on a Bayesian model, but the model chosen to yield inferences that are well-calibrated in a frequentist sense. Specifically, the aim is for posterior credible intervals that have (approximately) nominal frequentist coverage. To that end, models should be chosen that incorporate the main features of the sample design. Specifically, from a modeling or prediction perspective, the design weight is a variable known for all units in the population, and should be treated as a covariate in the prediction model (e.g., Gelman 2007); clustering and multistage sampling can be incorporated using hierarchical random-effects models. Two final examples are provided to illustrate the approach.

Example 13.6. Post-stratification on a categorical covariate

Consider an equal probability sample with a single categorical post-stratifying variable Z, for which known population counts N_h are available for each post-stratum h, $h = 1, 2, \ldots, H$. Let $\bar{y}_h$ be the sample mean in post-stratum h, based on sample size n_h, and $n = \sum_{h=1}^{H} n_h$, $N = \sum_{h=1}^{H} N_h$. The standard estimate of the population mean is the post-stratified mean:

$$\bar{y}_{\mathrm{PS}} = \sum_{h=1}^{H} P_h \bar{y}_h, \qquad (13.12)$$

where $P_h = N_h / N$ and N is the population size. This can be viewed as a weighted mean

$$\bar{y}_{\mathrm{PS}} = \sum_{h=1}^{H} \sum_{i=1}^{n_h} w_i y_i,$$

where $w_i = N_h / (N n_h)$ is the post-stratification weight for sampled units in poststratum h. These weights can be very large in post-strata with small sample sizes n_h, which, unlike stratified sampling based on Z, are not under the control of the sampler. These large weights can lead to excessive variability in $\bar{y}_{\mathrm{PS}}$. In fact, strictly speaking, $\bar{y}_{\mathrm{PS}}$ does not have a distribution in repeated sampling, because with positive probability

the sample sizes in some post-strata may be zero. This remains true if
the post-strata are modified to ensure that the post-strata sample counts
are all positive for the observed sample, for example, by pooling adjacent
strata.

The standard design-based approach to excessive variability of $\bar{y}_{\text{PS}}$ is
to modify the weights $\{w_i\}$, for example, by trimming the large ones.
However, from a prediction perspective, this is misguided. The problem
is not the weights – the population proportions $\{P_h\}$ in each post-stratum
are known, after all – the problem is that the estimates $\{\bar{y}_h\}$ have low
precision in post-strata with small sample sizes; indeed, the estimates do
not even exist in post-strata with no sampled units. It is the estimates in
sparse post-strata that need to be modified, not the weights $\{w_i\}$ attached
to sampled units. The principled way to modify $\{\bar{y}_h\}$ is to assume a model
relating Y and Z. One approach is to assume the normal random-effects
model

$$p(\mu_j \mid \mu, \tau, \sigma) \sim_{\text{iid}} G(\mu, \tau^2),\ p(\mu, \sigma, \tau) \propto \sigma^{-1}, \qquad (13.13)$$

which treats the post-stratum means as random effects. The variances in
each post-stratum might also be treated as distinct random effects and
assigned a prior distribution, rather than pooled. The posterior mean of
$\bar{Y}$ for the prior distribution (13.13) moves the weight w_i of sampled units
in post-stratum j toward one, with a degree of shrinkage that depends on
the relative size of estimates of σ and τ (Lazzeroni & Little 1998).

The prior distribution (13.13) makes the non-trivial assumption that the
post-stratum means are exchangeable. It can be relaxed by restricting
the random-effects model to a subset of post-strata with small sample
counts, or the constant mean μ in (13.13) might be replaced by a regression on known post-stratum characteristics C_j, as in:

$$p(\mu_j \mid \beta_0, \beta_1, \tau, \sigma, c_j) \sim_{\text{iid}} G(\beta_0 + \beta_1 c_j, \tau^2),\ p(\beta_0, \beta_1, \sigma, \tau) \propto \sigma^{-1}, \quad (13.14)$$

which limits the exchangeability assumption to the errors in the regression of μ_j on c_j. For extensive generalizations of this basic example, see
Gelman and Little (1997), Elliott and Little (2000), Elliott (2007), Gelman
(2007) and Si et al. (2020).

Example 13.7. Regression estimator of the mean, given a population auxiliary variable

If the auxiliary variable Z in the previous example is continuous, a common way to incorporate it in the inference is via the regression estimate
of the mean:

$$q = \bar{y}_{\text{REG}} = \bar{y} + \hat{\beta}_1 (\bar{Z} - \bar{z}),$$

where $\hat{\beta}_1$ is the least squares estimate of the slope of Y on Z in the sample,
and $\bar{z}$ and $\bar{Z}$ are respectively the sample and population mean of Z. In a

simulation study of five real populations, Royall and Cumberland (1981, 1985) assess inferences centered at q, with (a) the standard design-based standard error based on simple random sampling, namely:

$$I_{.95D} = \bar{y}_{\text{reg}} \pm 1.96\hat{\text{se}}_D(\hat{\beta}), \quad \hat{\text{se}}_D(\hat{\beta}) = \sqrt{(1-f)s_{Y.Z}/n},$$

where $s_{Y.Z}^2$ is the sample residual variance and $f = n/N$ is the sampling fraction; and (b) the prediction standard error based on the normal linear regression model with constant variance, namely:

$$I_{.95M} = \bar{y}_{\text{reg}} \pm t_{.975, n-2} 1.96\hat{\text{se}}_M(\hat{\beta}),$$

$$\hat{\text{se}}_M(\hat{\beta}) = \hat{\text{se}}_D(\hat{\beta})\sqrt{\left(1+(\bar{z}-\bar{Z})^2\right)/(1-f)\left(\sum_{i=1}^{n}(z_i-\bar{z})^2/n\right)}.$$

The design-based confidence intervals $I_{.95D}$ exhibit very poor conditional confidence coverage when the observed $\bar{z}$ deviates substantially from $\bar{Z}$. The model-based confidence interval $I_{.95M}$ takes into account this lack of balance with respect to $\bar{z}$ but is vulnerable to model misspecification, specifically lack of linearity in the relationship between Y and Z or non-constant residual variance. Robust estimates of standard error, based on the sandwich estimator or the jackknife, yield intervals with better conditional coverage properties, although still sometimes deviating from nominal coverage levels. An alternative approach is Bayesian inference based on a more flexible model relating Y to Z, such as the penalized spline model:

$$\left(y_i \mid z_i, \beta, \sigma^2\right) \sim_{\text{ind}} G\left(\text{spline}(z_i, \beta), \sigma^2 z_i^\alpha\right), \quad \text{spline}(z, \beta) = \beta_0 + \sum_{j=1}^{p}\beta_j z_i^j + \sum_{\ell=1}^{m}\beta_{\ell+p}(z_i - \kappa_\ell)_+^p$$

$$\left(\beta_{l+p} \mid \tau\right) \underset{\text{iid}}{\sim} N(0, \tau^2), l = 1, ..., m; \quad p(\beta_0, ..., \beta_p, \alpha, \sigma, \tau) \propto 1/\sigma, 0 < \alpha < 2,$$

$$\tag{13.15}$$

where the constants $\kappa_1 < ... < \kappa_m$ are selected fixed knots, and $(u)_+^p = u^p$ if $u > 0$ and 0, otherwise (see, e.g., Ruppert, Wand & Carroll 2003). The parameter α allows for a variety of common forms of heteroskedasticity. The Bayesian standard errors then reflect imbalance in distribution of Z in the sample and population, and the flexibility of the model limits bias from model misspecification.

Suppose that the target quantity is not the population mean of Y, but the least squares slope of Y on Z in the population. A robust approach is to impute the non-sampled values of Y using the model (13.15), and then estimate the slope of Y on Z as the least squares slope estimated on the filled-in population data. Uncertainty can be propagated by multiple imputation (Rubin 1987), a method founded on Bayesian ideas. In this context, Little (2004) distinguishes between the "target model" that determines the target population quantity of interest, here the linear regression of Y on X, and the "working model" (13.15) that is the basis for inference, and is used to predict survey variables for the non-sampled and nonresponding units in the population. Distinguishing between these two models provides for a robust form of Bayesian survey inference. Szpiro, Rice and Lumley (2010) apply a similar idea in a superpopulation regression setting.

13.6 Conclusions

Survey sampling is for me a fascinating area of statistics, combining a key idea in statistical design, namely probability sampling, nitty-gritty practical issues of data collection, an emphasis on population quantities that have a real existence outside the boundaries of a simplified statistical model, and lofty philosophical disputes between advocates of model-based and design-based inference. The paper by Neyman (1934) is widely cited as a landmark in the development of survey sampling as a discipline, developing a key design idea, stratified sampling, and paving the way for future complex sampling designs. I have argued that survey sampling is a great field of application for the calibrated Bayes philosophy discussed in Chapter 6. Further research on that aspect should be fruitful in the era of "big data," where true probability samples are increasingly difficult to obtain.

13.7 Some Thought Questions on This Chapter

1. State the main contributions of Neyman's (1934) paper, in your view.

2. For probability samples, describe briefly the difference between (a) "design-based" or "randomization" inference and (b) "model-based" inference.

3. Review the concept of design consistency in Isaki and Fuller (1982), and the models in Firth and Bennett (1998) that lead to design-consistent estimates. Consider whether seeking models that are design-consistent is aligned with the Calibrated Bayesian approach to survey inference.

4. Neyman's paper is often cited as the basis for design-based inference, but does his paper consistently take a design-based perspective?

14

Randomized Clinical Trials and the Neyman/Rubin Causal Model

The papers:

Medical Research Council (1948). Streptomycin treatment of pulmonary tuberculosis: a Medical Research Council investigation. *Brit. Med. J.*, 2, 769–782.

Rubin, D. B. (1978). Bayesian inference for causal effects: the role of randomization. *Ann. Statist.*, 6, 1, 34–58.

Other reading:

Little, R. J. & Lewis, R. J. (2021). Estimands, estimators and estimates. *J. Amer. Med. Assoc.*, 326, 10, 967–968.

14.1 Introduction

In the last chapter, I discussed the benefits of randomization in the *selection* of units from a population into a sample. In this chapter, I discuss another important role of randomization, in the *allocation* of alternative treatments in a comparative trial. Randomized clinical trials (RCTs) comparing alternative medical interventions are perhaps the principal area of application, although the idea is important in any scientific area that involves comparisons of alternative treatments, including economics and other social sciences.

RCTs have put the practice of medicine on a sound scientific basis; prior to their invention, the choice of treatments was often based on expert opinion rather than reliable empirical evidence, and as a consequence, many supposed cures were worse than the disease. Often, the best approach was to avoid doctors altogether and hope that one's immune system was up to the task of fighting the ailment. In Section 14.2, I discuss two pioneers of RCTs, Austin Bradford Hill and Paul Meier, and provide examples of randomized and non-randomized trials that illustrate why RCTs are a gold standard for clinical studies.

Clinical trials have the goal of assessing the causal effects of treatments, and thus are a form of causal inference. Sections 14.3–14.5 consider three seminal ideas in causal inference, namely:

a. the Neyman/Rubin definition of causal effect (Section 14.3);
b. internal validity, confounding, and the role of randomized assignment in promoting internal validity (Section 14.4);
c. external validity, effect modification, and the consequences of non-random selection of study participants (Section 14.5).

The presentation follows the summary of these topics in Little and Lewis (2021), which is included here as other reading.

The chapter concludes in Section 14.6 by summarizing the seminal paper by Rubin (1978), which provides a conceptual framework for the analysis of studies comparing treatments and a justification of the important role played by randomization in providing for robust statistical inferences. Although couched in terms of Bayesian models, it is also directly relevant to frequentist inference based on statistical models.

Rubin's paper incorporates the Neyman/Rubin definition of causal effects, the topic of Section 14.3. It also reveals the value of randomization by including indicators for selection and allocation as a part of the model. Randomization in the allocation of treatments ensures that the assignment mechanism is ignorable, which means that it does not need to be modeled, greatly simplifying the analysis. This parallels the role of random selection of individuals in sample surveys, which (as discussed in Chapter 13) makes the selection mechanism ignorable.

In my view, the conceptual simplicity and all-encompassing framework of causal modeling makes Rubin (1978) one of the most significant statistical papers of the 20th century.

14.2 Randomized Clinical Trials

Sir Austin Bradford Hill is widely considered the father of RCTs. He was the lead statistician in a pioneering RCT on streptomycin for the treatment of pulmonary tuberculosis (Medical Research Council 1948). Interestingly, Neuhauser and Diaz (2004) describe an earlier application of randomization to a tuberculosis trial conducted in Detroit, reported in Amberson, McMahon and Pinner (1931).

Chalmers (2011) observes that, prior to randomization, the main device for ensuring comparability of treatment groups in clinical trials was alternation, where assignment was alternated between the compared treatments. If those assigning the treatments knew the alternation scheme, the assignment could be manipulated (consciously or unconsciously) to favor a treatment. Chalmers writes:

> In an internal report for the MRC dated 22 December 1933, Hill expressed concern about the allocation of patients to comparison groups in a MRC study of serum treatment for pneumonia in which alternation should have been used. Imbalance in the sizes of the comparison groups made clear that alternation had not been strictly observed, prompting Hill to stress in his memorandum that greater effort should be taken 'that the division of cases really did ensure a random selection'. In other words, to control allocation bias successfully, Hill realized that it is crucially important to conceal the allocation schedule from those involved in entering participants, thus preventing foreknowledge of allocations.

Paul Meier was a fierce advocate of RCTs in the United States. His obituary in the Daily Telegraph read in part:

> Paul Meier… was a statistician who championed the idea of testing new medical treatments through randomised trials, so helping to lead a revolution in clinical research and saving, albeit indirectly, millions of lives…… The idea of assigning subjects in medical trials solely on the basis of random selection might now seem obvious. But, like many medical innovations, it did not seem so at the time Meier proposed it in the 1950s…. Many physicians were horrified at the idea that their selection should be random, together with an equally randomly-selected "control" group of patients who were given the standard treatment or a placebo… At first Meier's arguments met with incomprehension: "When I said 'randomise' in breast cancer trials," he recalled in 2004, "I was looked at with amazement by my medical colleagues: 'Randomise? We know that this treatment is better than that one.' I said, 'Not really!'

Recently, the world has suffered through the Covid-19 pandemic, at a cost of millions of lives. Many were saved by the invention of mRNA vaccines, which are remarkably effective in staving off the worst effects of the disease. Why am I so sure that the vaccines work? It is the product of remarkable science, but the empirical evidence comes from large RCTs like those in Example 14.1, which establish beyond reasonable doubt that the vaccines are effective.

**Example 14.1. An RCT to assess the efficacy and safety
of an mRNA vaccine for Covid-19**

Polack et al. (2020) conducted a large RCT to assess the effectiveness of BNT162b2 mRNA Covid-19 vaccine. The abstract reads, in part:

> In an ongoing multinational, placebo-controlled, observer-blinded, pivotal efficacy trial, we randomly assigned persons 16 years of age or older in a 1:1 ratio to receive two doses, 21 days apart, of either placebo or the BNT162b2 vaccine candidate (30 µg per dose)… The

primary end points were efficacy of the vaccine against laboratory-confirmed Covid-19 and safety... A total of 43,548 participants underwent randomization, of whom 43,448 received injections: 21,720 with BNT162b2 and 21,728 with placebo.

There were 8 cases of Covid-19 with onset at least 7 days after the second dose among participants assigned to receive BNT162b2 and 162 cases among those assigned to placebo; BNT162b2 was 95% effective in preventing Covid-19 (95% credible interval, 90.3 to 97.6). Similar vaccine efficacy (generally 90 to 100%) was observed across subgroups defined by age, sex, race, ethnicity, baseline body-mass index, and the presence of coexisting conditions. Among 10 cases of severe Covid-19 with onset after the first dose, 9 occurred in placebo recipients and 1 in a BNT162b2 recipient. The safety profile of BNT162b2 was characterized by short-term, mild-to-moderate pain at the injection site, fatigue, and headache. The incidence of serious adverse events was low and was similar in the vaccine and placebo groups.

No clinical studies are perfect, and many have conflicting conclusions, as in the following example:

Example 14.2. The treatment of advanced cancer by Vitamin C

Linus Pauling, the famous scientist and Nobel Prize winner, was a huge believer in the positive benefits of Vitamin C. His paper with Cameron (Cameron & Pauling 1976) purports to show large positive effects of Vitamin C in the treatment of advanced cancers. An excerpt from the abstract states:

The results of a clinical trial are presented in which 100 terminal cancer patients were given supplemental ascorbate as part of their routine management. Their progress is compared to that of 1000 similar patients treated identically, but who received no supplemental ascorbate...The mean survival time is more than 4.2 times as great for the ascorbate subjects (more than 210 days) as for the controls (50 days)... The results clearly indicate that this simple and safe form of medication is of definite value in the treatment of patients with advanced cancer.

The RCT reported in Creagan et al. (1979) reaches a completely different conclusion:

150 patients with advanced cancer participated in a controlled double blind study to evaluate the effects of high-dose vitamin C on symptoms and survival... Patients were divided randomly into a group that received Vitamin C (10 g per day) and one that received a comparatively flavored lactose placebo. 60 evaluable patients received vitamin C and 63 received a placebo.... In this selected group of patients, we were unable to show a therapeutic benefit of high-dose vitamin C.

Why do these two studies reach such different answers? Not because of sampling variability – the effects reported in Cameron and Pauling (1976) are much too large to be attributable to random variation. It is the design of the studies – one randomized, one not – that is the main factor. Later studies replicated the results in Creagan et al. (1979), and Vitamin C is not considered to be a useful treatment for cancer.

**Example 14.3. Vaccines and autism. Poorly designed
studies can do great harm**

The small (12 children) and methodologically flawed nonrandomized study by Wakefield et al. (1998) was later retracted because of selection biases, conflict of interest concerns, and ethical violations. Nevertheless, it was widely publicized and became a primary source for the anti-vaxing movement. A website on science-based medicine, https://www. sciencebasedmedicine.org/reference/vaccines-and-autism/

writes:

> In recent years the antivaccine movement has focused on the claim that vaccines are linked to neurological injury, and specifically to the neurological disorder autism, now referred to as autism spectrum disorder (ASD). However the scientific evidence overwhelmingly shows no correlation between vaccines in general, the MMR vaccine specifically, or thimerosal (a mercury-based preservative) in vaccines with ASD or other neurodevelopmental disorders.

Vaccine hesitancy has led to measles outbreaks and lengthened the COVID-19 pandemic, at a cost of many lives.

14.3 The Definition of Causal Effects

14.3.1 The Neyman/Rubin Causal Model

When is a treatment effect causal? That is, how do we know that better outcomes are caused by the treatment and not some other factor? How do we define a causal effect? Phenomena have multiple causes, often too hard to disentangle. Consider, for example, the hot-button topic of mass shootings. Shootings have multiple causes – ready access to guns, lack of gun training, security lapses, mental health of shooters, and so on. Individuals on different sides of the political spectrum can choose which of these causes to emphasize. But we can define the average *causal effect* of a specific policy to reduce mass shootings, such as the reduction of annual incidence of deaths by mass shootings, and this is what really matters in policy discussions.

The Neyman/Rubin Causal Model defines the causal effect of treatment for an individual as the difference in outcome under active treatment and under control (Rubin 1974). From this perspective, the estimation of causal effects is basically a missing data problem, because we only get to see the outcome from one treatment, the treatment received. However, we can estimate average causal effects in groups, and here, randomized treatment assignment plays a key role in avoiding bias. The following hypothetical example may help to make these ideas more concrete.

**Example 14.4. The causal effects for a hypothetical study
of two treatments for depression**

Figure 14.1 provides a simple numerical illustration of the definition,
for a trial of two alternative treatments of depression, a control treatment 1 and a new treatment 2. Subjects 1 and 2 are given treatment 1,
and subjects 3 and 4 are given treatment 2. The outcome is a depression score Y, and $Y(j)$ is the depression score given treatment j, where
higher scores mean more depressed. Figure 14.1(A) shows the data
that are actually observed, and Figure 14.1(B) shows the hypothetical data if we were actually able to observe the outcomes under both
treatments. The numbers are made up so that the average outcomes
based on outcomes actually observed (indicated by the asterisk) are
10 for treatment 1 and 9 for treatment 2, so the naïve estimate of the
average causal effect of treatment 2 is $9 - 10 = -1$, suggesting that
treatment 2 is better than control. (We ignore the small sample size
and statistical uncertainty here.) However, based on the hypothetical data in Figure 14.1(B), the causal effect of treatment 2 is positive
for all the subjects, suggesting that the new treatment has a negative
effect on depression.

14.3.2 The Importance of a Comparator Treatment

A central feature of the definition of causal effect in Section 14.3 is that it involves
a comparison of two possible outcomes. This implies that a good study design
for assessing the effectiveness of a treatment requires a comparator "control"
treatment – either a placebo treatment or an alternative active treatment with
which the new treatment can be compared. Unfortunately, many clinical studies do not have this feature and have what I like to call the "snake-oil salesman" (SOS) design. Namely, they compare a measure of the target ailment
before and after administration of the treatment, and see whether participants
got better.

Subject	Y(1)	Y(2)	Y(2)-Y(1)
1		6	
2		12	
3	9		
4	11		
Mean	10*	9*	-1*

(a)

Subject	Y(1)		Y(2)		Y(2)-Y(1)	
1	[1]	6	6		[5]	
2	[3]	12	12		[9]	
3	9	9	[10]		[1]	
4	11	11	[12]		[1]	
Mean	10*	[6]	9*	[10]	-1*	[4]

(b)

FIGURE 14.1
A simple numerical illustration of the Neyman/Rubin definition of causal effects. $Y(j)$ – depression score given treatment j, $j = 1, 2$. Means with an asterisk (*) are computed over the observed
data. (a) Observed data. (b) Underlying full data.

Do you suffer from a chronic disease, like depression? The salesman has a miraculous snake oil – take the recommended dose and your depression will go away! He cites a clinical study, where a set of individuals experiencing depression were given the oil, and their average depression score dropped from 8.6 to 6.3 after taking the medicine.

Does this constitute strong evidence that the medicine works? With a chronic disease like depression with a fluctuating course, if you select individuals with high levels of depression and give them a treatment, the average level of depression will decrease even if the treatment has no effect because of regression to the mean and potential placebo effects. So, the implication that the measure of disease will be unchanged under the absence of treatment is unjustified. Real evidence requires a comparator treatment.

SOS designs are common in medical studies, and in particular in studies of arthroscopic debridement, a once-popular surgical treatment of degenerative joint disease. Sprague (1981) summarized the findings of one such study as follows:

> **Example 14.5. A snake oil salesman (SOS) design: the treatment of degenerative knee joint disease by arthroscopic debridement**
>
> A series of 77 knees in 72 patients, ages ranging from 24 to 78 years (mean, 56 years), with moderate or severe degenerative arthritis were treated by percutaneous debridement of the joint under arthroscopic visualization.... Sixty-two patients with 68 knees were followed for at least six months, with a mean follow-up of 13.6 months. Subjectively, 84% of the patients were found to have a good or fair result. Complications were few and mild in nature, and there was little morbidity. Arthroscopic debridement of the knee joint is recommended as a useful therapeutic modality in many patients with degenerative arthritis of the knee.

Note the absence of a comparator, and the potential for regression to the mean, and placebo effects given the subjective outcome. A number of other studies follow a similar design. However, later RCTs (e.g., Moseley et al. 2002) compared this treatment with a sham surgery, where individuals were anesthetized but did not receive the surgery. These studies showed no advantage of debridement over the sham surgery.

14.4 Confounding and Internal Validity

An estimated average causal effect of a treatment has *internal validity* if it is a valid estimate of the average causal effect for the individuals included in the study. An estimated average causal effect of a treatment has *external validity* if it is a valid estimate of the average effect for the individuals who comprise the target population for the treatment. Clearly, external validity is a stronger requirement than internal validity, and internal validity is a prerequisite for external validity. In this section, we focus on the requirements for internal validity; external validity is the topic of the next section.

The key condition for internal validity is the absence of *confounding variables*.

Definition 14.1. A variable Z is a confounder if:

    a. it is a pre-treatment variable, not a consequence of the treatment;
    b. Z has a different distribution for participants in the treatment groups;
    c. Z is related to the outcome measure of interest.

If a study has confounding variables, the observed treatment effect may be attributable to the effects of these variables rather than the treatments per se. That is, confounding compromises internal validity. Thus, confounding variables can give rise to the kind of bias in the treatment effect estimated from the observed means in Figure 14.1(A).

The key benefit of randomization in the allocation in treatments is that the randomization ensures that, on the average, any potential confounding variables will have the same distribution across treatment groups. That is, condition (b) for a confounding variable is not satisfied. So, randomization ensures an *unconfounded assignment mechanism*. That is why RCTs are the gold standard in clinical research.

Formally, let t_i be an indicator for treatment assignment, taking value j if a unit i is assigned to treatment j. If the assignment is random, then the assignment does not depend on the outcomes under either treatment, that is:

$$\Pr\big(t_i = j \,|\, y_i(1), y_i(2)\big) = \Pr(t_i = j). \tag{14.1}$$

This implies that the distribution of potential outcomes does not depend on the assignment indicator, that is:

$$f\big(y_i(1), y_i(2) \,|\, t_i = j\big) = f\big(y_i(1), y_i(2)\big). \tag{14.2}$$

So, the mean outcome for each treatment j in the whole sample can be estimated by the conditional mean outcome given assignment to that treatment. More generally, in designs with stratified random assignment, the distribution of the assignment indicator is allowed to depend on the values of baseline covariates, say z_i. In that case, the two equations apply conditional on z_i.

RCTs have limitations, as discussed in the next section, and they are not always feasible: in assessing the potential health effects of a chemical pollutant, individuals cannot be randomized to exposure or non-exposure to the pollutant, or in assessing the health effects of smoking, people cannot be randomized to smoke or not to smoke. In such cases, evidence must come from observational studies, where assignment is not randomized. A key issue with observational studies is the possibility of confounding variables, as in the following example.

Example 14.6. Confounder bias in learning health systems

An administrative health system captures data for 200 patients with a rare disorder – 100 took Drug A and 100 Drug B. 70 people taking Drug A are "cured," and 30 people taking Drug B are "cured." The naïve conclusion is that Drug A is more effective – the difference is too large to be attributable to chance. But we can't conclude that Drug A is better – maybe something other than the effect of the drug – a confounding factor – explains the difference.

Statistical methods such as regression or propensity score methods, as discussed in Chapter 15, can correct for potential confounders that are measured, but cannot adjust for unobserved confounders in the analysis. So, a key feature of a good clinical database is that all potential confounding variables are recorded.

The randomization in a randomized study ensures that (at least in large samples) the distributions of both observed *and unobserved* confounders are the same across treatment groups. Chance imbalances in observed confounders can be handled by regression, and chance imbalances in unobserved confounders are reflected in the standard error of the estimated treatment effect.

14.5 Effect Modification and Its Impact on External Validity

The control of confounding and consequent internal validity is the main reason why RCTs are the gold standard in clinical studies. However, the participants in RCTs are rarely if ever randomly sampled from the target population, and usually consist of volunteers who consent to participate. Thus, RCTs are subject to the bias in the selection of participants discussed in Chapter 13. The potential for selection bias does not affect internal validity, where inferences are restricted to individuals participating the trial, but it does affect external validity, which involves extrapolating the results of the study to the target population.

If the causal effect of the treatment was identical for all individuals in the population, then nonrandom selection of individuals into the study does not lead to bias. If, however, the effect of the treatment varies for different individuals, a feature known as *effect modification*, then the average treatment effect estimated from selected participants may be a biased estimate of the average treatment effect in the population. So, the key issue for external validity is the degree of effect modification, and this is not ensured by random assignment of treatments.

Statistically, effect modification is a form of interaction between the treatment indicator and a baseline variable, such as age or stage of disease:

Definition 14.2. A variable Z is an effect modifier if:

a. it is a pre-treatment variable, not a consequence of the treatment; and
b. the variable is associated with the treatment effect, that is, the treatment effect varies in subpopulations defined by values of Z.

If Z is a pre-treatment variable, Y(t) is the outcome given treatment t, and the treatment effect is measured as a difference in means, then Z is *not* an effect modifier if:

$$E(Y(2) - Y(1) \mid Z) = E(Y(2) - Y(1)),$$

for all Z.

Effect modification can be assessed for baseline variables that are measured in the study, by comparing the size of treatment effect in subgroups defined by those variables. If the treatment effect is similar across these subgroups, then the evidence for external validity is enhanced. For this reason, many clinical trials compare treatment effects across subgroups defined by baseline variables. In particular, in Example 14.1, the abstract notes that the efficacy of the vaccine appears similar across baseline characteristics. Such analyses are useful, but it is important to note that effect modification is a form of interaction, and the power to assess interactions is much lower than the power to detect main effects. Clinical trials are much more expensive than the analysis of existing clinical databases, so the sample sizes are generally much smaller. So, the power to detect effect modification is very limited, and of course restricted to effect modifiers that are measured in the study.

Estimates of treatment effects from RCTs are often strong in terms of internal validity but potentially weak in terms of external validity and for assessing effect modification. Estimates of treatment effects from large clinical databases are often weak in terms of internal validity but can supplement RCTs by limiting the effects of selection bias and having large sample sizes. So, these two data types have different strengths and weaknesses, and the future may lie in studies that combine results from clinical trials and databases, as discussed, for example, in Walicke et al. (2017).

14.6 The Rubin (1978) Framework for the Analysis of Studies of Causal Effects

A common criticism of Bayesian methods is that they do not appear to rely on randomization, either in the selection of units or the allocation of treatments, because inferences are based on a model for the study variables, rather than the randomization distributions for selection (in the case of survey sampling) or treatment allocation (in the case of comparative trials). Rubin (1978) argues against this position, writing:

> Some opponents of randomization turn to Bayesian statistics as a conceptual foundation for their position. However, careful development of the Bayesian framework for drawing inferences about causal effects of treatments explicates the steps required to analyze randomized and nonrandomized studies, and demonstrates that randomized studies are in general substantially easier to analyze than comparable nonrandomized studies. Therefore, we argue that randomization plays a central role in Bayesian inference for causal effects.

Figure 14.2 from Rubin (1987) depicts causal inference as a giant missing data problem. The rows are individuals in the population, a sample of which is selected for the study. The columns consist of

a. baseline variables X, measured prior to the assignment of treatments;

b. a variable W indicating which treatment is assigned, taking value 0 for individuals in the population not included in the study, and t for those included and assigned treatment t, for $t = 1,..., T$.

c. T collections of d post-treatment variables $Y^t = \{(Y_1^t,...,Y_d^t), t = 1,...,T\}$, representing the set of values if that unit was assigned to each treatment t, under the Neyman/Rubin causal model of Section 14.3; and

d. missing data indicators M representing values that are not recorded. These might include values that should be measured according to the study protocol but are not recorded because of missing data.

Conceptually, the modeling task is to fill in (that is, predict) the unrecorded values in the matrix, after which summary estimates of treatment effects can be computed. Note that very few of the values in the matrix are actually recorded – at most, the values of pre-treatment variables, and post-treatment variables for the treatment assigned for the included individuals. Multiple imputation (MI, Rubin 1987) provides a tool for predicting the missing values that allows the uncertainty of estimates to be measured, using MI combining rules. In MI,

Pretreatment values	Which treatment	Posttreatment values		Missing data indicator		
X	W	Y		M		
		Y^1 …	Y^T	M^X	M^1 …	M^T
X_1 … X_c	W	Y_1^1 … Y_d^1 …	Y_1^T … Y_d^T	M_1^X … M_c^X	M_1^1 … M_d^1 …	M_1^T … M_d^T

Experimental units in population P: $1, 2, \ldots, N$

FIGURE 14.2
All values in a study of T treatments. (*Source*: Figure 14.1 in Rubin (1978), reprinted with the permission of the Institute of Mathematical Statistics.)

multiple data sets are created with different draws from the distribution of the missing values under a statistical model. Causal effects are estimated from each filled-in data set, and the results are combined using simple MI combining rules.

The key feature of Rubin (1978) is that the indicators W for assignment and M for missingness are treated as random variables along with X and Y. Thus conceptually, Rubin considers a Bayesian model for the joint distribution of (X, Y, W, M) indexed by parameters θ, which are assigned a prior distribution $\pi(\theta)$ in a Bayesian analysis. Writing $(\tilde{X}_{(1)}, \tilde{Y}_{(1)})$ for the observed values and $(X_{(0)}, Y_{(0)})$ for the unobserved values of (X, Y), $\tilde{X} = (X_{(0)}, \tilde{X}_{(1)})$, $\tilde{Y} = (Y_{(0)}, \tilde{Y}_{(1)})$ and $(\tilde{W}, \tilde{M})$ for the observed values of (W, M), he then expresses the posterior predictive distribution of $Y_{(0)}$ as

$$\Pr(\tilde{Y}_{(0)} \mid \tilde{X}_{(1)}, \tilde{Y}_{(1)}, \tilde{W}, \tilde{M})$$

$$= \frac{\iint f_{W|XY}(\tilde{W} \mid \tilde{X}, \tilde{Y}, \pi) f_{M|WXY}(\tilde{M} \mid \tilde{X}, \tilde{Y}, \tilde{W}, \pi) f_{XY}(\tilde{X}, \tilde{Y} \mid \pi) p(\pi) d\pi dX_{(0)}}{\iiint f_{W|XY}(\tilde{W} \mid \tilde{X}, \tilde{Y}, \pi) f_{M|WXY}(\tilde{M} \mid \tilde{X}, \tilde{Y}, \tilde{W}, \pi) f_{XY}(\tilde{X}, \tilde{Y} \mid \pi) p(\pi) d\pi dX_{(0)} dY_{(0)}},$$

integrating over the missing values of all the variables.

In the clinical trial context, it is useful to replace W by (S, A), where S is the binary variable taking value 1 if a unit is selected and 0 otherwise, and A is the treatment assignment for selected units. This leads to a joint distribution of (X,Y,S,A,M), which can be factored as:

$$f(X,Y,S,A,M\,|\,\theta)$$
$$= f_{XY}(X,Y\,|\,\theta)f_{S|XY}(S\,|\,X,Y,\theta)f_{A|SXY}(A\,|\,S,X,Y,\theta)f_{M|SAXY}(M\,|\,S,A,X,Y,\theta).$$

The specification of this joint distribution, in a form where the parameters are estimable, is a very challenging modeling task. Rubin's basic idea is that if the distributions of the indicators S, A, or M can be assumed ignorable, in the sense that they depend only on observed data, then they do not have to be included as random variables in the model, thus simplifying the modeling task. Sufficient conditions for ignorability of M were described in Rubin (1976) and discussed in Chapter 13, and similar conditions apply for ignorability of S and A.

Specifically, if selection S is assumed ignorable, a model is only required for (X,Y,A,M); if selection S and assignment A are assumed ignorable, a model is only required for (X,Y,M); and if selection S, assignment A, and missingness M are assumed ignorable, a model is only required for (X,Y).

Selection S would be ignorable if the participants in a clinical trial were randomly sampled from the target population. In practice this very rarely happens, so ignorability of S is a strong assumption. An alternative to this assumption is to condition the model on $S = 1$, that is, on selected cases. This leads to an assessment of internal validity but not of external validity. Assignment A is ignorable in a randomized trial but is an assumption in an observational study, and finally, missingness M is not under control of the trialist, so ignorability of M is an assumption. Thus, we see how Rubin's ideas neatly tie together the major threads in this and the previous chapter.

14.7 Some Thought Questions on This Chapter

1. Ads on television often promote products for reducing weight, increasing energy levels, or some other health benefit. In the light of the concepts in this chapter, consider whether these ads provide evidence of a genuine causal effect of the product.

2. Distinguish between the cause of an ailment (e.g., asthma) and the causal effect of a treatment for the ailment.

3. Find two research studies on a medical topic of interest to you and assess their relative strengths and weaknesses in terms of statistical study design.

4. Explain in words the difference between a confounder and an effect modifier.

5. Does randomization in clinical trials help ensure internal or external validity? Explain why.

6. In RCTs, the distribution of participants across treatments are often compared in subgroups of the sample defined by pre-treatment variables like age, gender, and disease stage. What do these displays tell us about internal or external validity?

7. In RCTs, treatment effects are often compared in subgroups of the sample defined by pre-treatment variables like age, gender, and disease stage. What do these displays tell us about internal or external validity?

8. Review principal stratification (Frangakis and Rubin 2002), an approach to the analysis of post-treatment confounders based on the Neyman/Rubin definition of causal effects.

15

Propensity Score Methods

The paper:

Rosenbaum, P. R. & Rubin, D. B. (1983). The central role of the propensity score in observational studies for causal effects. *Biometrika*, 70, 1, 41–55.

Other readings:

Little, R. J. (2022). Some reflections on Rosenbaum and Rubin's propensity score paper. *Observational Studies*, 9, 1, 69–75.

15.1 Introduction

In Chapter 14, I discussed the value of randomized assignment in studies comparing treatments. However, randomization is not always feasible or ethical, and observational studies play an important role when randomized studies are not available. When designing and analyzing observational studies, it is useful to mimic the corresponding design and analysis for a randomized trial. In that context, the propensity score is the analog of the probability of assignment in randomized trials, and conditioning on an estimate of the propensity score is a useful way of controlling for confounders in observational studies.

The key idea of conditioning on the estimated propensity score is in the seminal paper by Rosenbaum and Rubin (1983), henceforth RR, the main paper in this chapter. The basic idea is tantalizingly simple – my reaction is "why didn't I think of that?" Also, not just one but three important applied problems – bias from the assignment of treatments, selection bias, and bias from nonresponse – can be addressed by estimating and adjusting for the propensity of these quantities. The underlying framework is laid out in Rubin (1978), which was discussed in Chapter 14.

In Section 15.2, I review the main ideas in RR, and why the method provides a useful alternative to classical regression approaches. The standard approach to adjustment for a set of observed potential confounders X is multiple regression of the outcome Y on indicators for treatment (say W) and X. This approach is vulnerable to misspecification of the regression model, in ways that may be hard to detect with a large number of predictors.

RR provides robustness by reducing the covariates to a single variable, the estimated propensity to be assigned treatment given X. This reduction facilitates the injection of robustness into the analysis, by methods that allow for a flexible relationship between the propensity score and study outcomes.

In Section 15.3, I discuss how the methods in Section 15.2 can be adapted to adjust for nonresponse and selection bias. One of the methods discussed there is a spline-based regression method, penalized spline of propensity prediction (PSPP). In Section 15.4, I return to the problem of adjusting for confounders in treatment assignment, describing the analog of PSPP for that setting, penalized spline of propensity for treatment comparisons (PENCOMP).

My description of these ideas is based on Little (2022), which is included as other reading.

15.2 Rosenbaum and Rubin (1983)

15.2.1 Key Properties of the Propensity Score

As discussed in Chapter 14, confounder bias is a key concern in the analysis of observational studies for comparing treatments. A confounding variable is a pre-treatment variable that is (a) related to the outcome measure Y, and (b) has a distribution that differs across treatment groups. The objective of RR is to assess the causal effects of treatments in observational studies, given observed potential confounders X. A key assumption is that there are no unmeasured confounders.

For simplicity, consider the case of just two treatments, say a new treatment and a control treatment, and let W denote the treatment assignment, with $W = 1$ for the new treatment and $W = 0$ for the control treatment. Values of W and X are measured for the set of units in the study. The propensity to be assigned $W = 1$ is estimated by regressing W on X, using a method such as probit or logistic regression appropriate for a binary outcome. The propensity score for each unit is then the predicted probability that $W = 1$ given X from this regression. The treatment effect is then estimated adjusting for the estimated propensity score. Four approaches to adjustment – stratification, matching, weighting, and regression – are discussed below.

The neat feature of the method is that adjustment for a single variable – the estimated propensity score – simultaneously adjusts for all the confounding variables X, thus reducing the problem to adjusting for a single variable. This is a consequence of the fact that the propensity score is a *balancing* score, which means that conditional on the propensity, the distribution of all the confounding variables X is the same for the two treatments. Thus, condition (b) above for being a confounder no longer applies for the variables X after adjustment for the propensity score.

The mathematics underlying the approach is remarkably straightforward. Let $\Pr(W = 1 | X)$ denote the propensity to be assigned treatment 1 given X, and let Y measure the treatment outcome. Assume ignorable treatment assignment, that is, no unmeasured confounders:

$$\Pr(W = 1 | X, Y) = \Pr(W = 1 | X), \tag{15.1}$$

and also assume that this probability is positive for all values of X – the *positivity* assumption, discussed more later. Then

$$\Pr\left(W = 1 | Y, \Pr(W = 1 | X)\right) = E_X\left[\Pr(W = 1 | Y, X) | Y, \Pr(W = 1 | X)\right]$$

$$= E_X\left[\Pr(W = 1 | X) | Y, \Pr(W = 1 | X)\right]$$

by Eq. (15.1)

$$= \Pr(W = 1 | X).$$

So, W is independent of Y given $\Pr(W = 1 | X)$, and conditioning on $\Pr(W = 1 | X)$ removes the selection bias. Also, because $\Pr(W = 1 | X)$ is a function of X:

$$\Pr\left(W = 1 | X, \Pr(W = 1, X)\right) = \Pr(W = 1 | X),$$

which implies that W and X are independent given the propensity $\Pr(W = 1 | X)$. Hence, $\Pr(W = 1 | X)$ is a balancing score, and after conditioning on it, the variables X are no longer confounders. RR also show that the propensity is the coarsest function of X that is a balancing score.

In practice, the propensity score is not known and has to be estimated from the regression of W on X. The above properties hold if this regression is well-specified. One might wonder whether the resulting method is an improvement over the classical method of multiple regression of Y on W and X, because it has just replaced the problem of correctly specifying one multiple regression, the regression of Y on W and X, by another, the regression of W on X. However, the balancing property of the propensity score is easily checked, by comparing distributions of observed variables within estimated propensity score categories. If lack of balance remains, the regression of W on X can be modified to address it. Having achieved balance, the propensity score method requires adjustment for a single variable, the propensity to be assigned, rather than a set of variables, X. Multiple regression of Y on X makes more assumptions and is harder to specify correctly than adjustment of the estimated propensity score.

Even if the regression of W on X is not correctly specified, the lack of balance will often be greatly reduced by conditioning on the estimated propensity, thus greatly reducing confounder bias even if it is not entirely eliminated.

15.2.2 Methods of Adjustment for the Propensity Score

There are four main approaches to adjusting for the propensity score after it has been estimated. The simplest is stratification, where strata are formed based on grouping the estimated propensity into similar values, and the causal effect of treatment is estimated within each stratum. These estimates are then combined to obtain an overall estimated causal effect.

Another approach to adjustment is to match individuals assigned to treatment and control that have similar propensities to be assigned treatment. Individuals who do not have close matches are dropped from the analysis. This method is not necessarily efficient, and matching based on other covariates as well as the propensity may improve efficiency (see, e.g., Greifer & Stuart 2021).

Inverse probability weighting weights units by the inverse of their estimated probability of assignment. Like matching, this approach can be inefficient, particularly when the propensities of assignment involve variables that are not strongly related to the outcome. Augmented inverse probability weighting predicts the outcome based on a regression on X and then weights residuals from the regression by the inverse of the estimated probability of assignment. This is more efficient than weighting alone and has a double robustness (DR) property of giving valid estimates of treatment effect if either the prediction model for Y or the propensity model is correctly specified.

The final method is to predict the outcomes for treatments not assigned via a regression model, including the estimated propensity score as a covariate. This approach has the advantage that, unlike weighting, the regression adjustment takes into account strength of relationship between propensity and outcome; for example, if the propensity is assumed linearly related to the outcome, then the magnitude of the slope of the regression coefficient is small if the relationship is weak and increases as the strength of the relationship increases.

Regression avoids inefficiencies arising from weighting when the weights are highly variable. However, a linear regression can lead to bias if the shape of the relationship between the outcome and the propensity is not correctly specified. This can be addressed by specifying a flexible relationship between the outcome and the response propensity, such as a spline. This motivates penalized spline for treatment comparisons (PENCOMP), which I discuss in Section 15.4. Before that, I review the use of propensity methods in the context of nonresponse and selection bias.

15.3 Propensity Score Methods for Handling Nonresponse

15.3.1 Response Propensity Stratification

David et al. (1983) applies the propensity score idea in RR to survey nonresponse rather than treatment assignment. A canonical problem is unit nonresponse in surveys, where data are available for respondents on a set of survey outcomes $Y_1, ..., Y_p$ and on both respondents and nonrespondents for a set of variables $X = (X_1, ..., X_k)$. In some settings, values of some components of X are available for the whole population, but I do not focus on this distinction here. Let R denote the response indicator, taking value 1 for respondents and 0 for nonrespondents. The propensity to respond is estimated by a regression of R on X, using a model such as logistic regression appropriate for a binary outcome. The estimate of the probability of response is computed for each respondent, and the inverse of this probability becomes a nonresponse weight. In *response propensity (RP) stratification*, categories of the estimated response probabilities are formed, and the nonresponse weight is the inverse of the sample response rate within these categories. This categorization can avoid very large weights by judicious choice of categories.

RP stratification can reduce bias due to X when it is related to the survey variable. However, when X is predictive of response (so that the weights are variable), but the response propensity is weakly related to the survey outcome, weighting reduces precision with no compensating reduction in bias (Little & Vartivarian 2005). One approach to reducing unnecessary variability in the weights is to restrict the predictors in response propensity model to variables that are predictive of the main survey outcomes. For an application of this idea, see Morral et al. (2014). Another approach is to simply drop respondent cases with low estimated response propensities and hence high weights, which may have undue influence on survey estimates. This improves precision but effectively restricts the estimand to a subpopulation of units that is more likely to respond.

15.3.2 Penalized Spline of Propensity Prediction (PSPP)

The standard approach to unit nonresponse adjustment uses the inverse of the estimated response propensity as a weight, as in Section 15.3.1. From the Bayesian prediction perspective, it is more natural to treat the estimated response propensity as a covariate in a model to predict survey nonrespondents. What are the pros and cons of weighting and prediction?

For a single categorical predictor, weighting by the inverse response rate in a cell is equivalent to prediction based on a regression of Y on the set of dummy variables for the cells. In other settings with continuous covariates, weighting and prediction differ. Suppose we estimate the propensity to respond by a regression of R on a set of covariates and compare estimates of the mean of a survey variable Y by (a) weighting respondents by the inverse of the estimated response propensity, or (b) predicting nonrespondent values of Y by linear regression on the estimated propensity to respond. Both approaches adjust appropriately for nonresponse bias due to the covariates, if the relevant models are well-specified.

The weighting approach is simpler with multiple Y's, because the weights are the same for all the variables Y. The prediction approach is more nuanced, because the regression coefficient of the estimated propensity accounts for the degree of association between the propensity and Y. In particular, as noted above, if the covariates X are strong predictors of the propensity but the propensity is a weak predictor of Y, weighting or prediction is not needed to adjust for bias, and the weighted mean has higher mean squared error than the unweighted mean, that is, weighting makes the estimator worse (Little & Vartivarian 2005). Prediction is more efficient than weighting, because when the association is weak, the small resulting coefficient of the regression on the propensity dampens the adjustment.

If the linear regression model on R is misspecified, then it doesn't completely eliminate bias in the estimated mean. In fact, the potential bias from misspecification is most serious when the method has the greatest potential, namely the predictors are strongly related to both nonresponse and the survey variable. This motivates PSPP Little & An 2004; Zhang & Little 2009), which regresses the survey variable on a penalized spline of the estimated propensity, thus allowing a more flexible relationship between the variables. Other types of splines could also be fitted, but the penalized spline is easily fit with readily available mixed-model software.

Suppose the propensity is a weak predictor of the survey variable, but a different combination of the covariates is a strong predictor. For example, there are two covariates X_1 and X_2, the logit of the propensity is linear in $X_1 + X_2$, and the best predictor of Y is $X_1 - X_2$. Weighting or prediction on the propensity are both inefficient here (weighting more so than prediction). Weighting can be improved by augmented inverse-probability weighting (AIPW), which regresses on the covariates and applies weights to the residuals. PSPP simply adds strong predictors as additional covariates in the regression model for the survey outcome, along with the penalized spline of the propensity. Including all the predictors leads to issues with multicollinearity, so one of the predictors typically needs to be dropped.

PSPP has a potential advantage in efficiency over AIPW in this setting – if the added covariates in PSPP yield close to best predictors of Y, then the residuals from the regression of Y on these covariates are weakly related to covariates and hence to the propensity. Weighting the residuals by the inverse propensity is less efficient than PSPP. In simulations, I have found PSPP to be similar to AIPW in terms of bias reduction, but potentially more efficient, particularly when the weights in AIPW are very variable (Yang & Little 2015; Zhang & Little 2011).

A feature of the PSPP model is that the regression coefficients in the propensity are estimated, and hence are subject to error. The Bayesian version of PSPP propagates this uncertainty by including a prior distribution for these unknown regression coefficients in the propensity model. An alternative approach is to multiply-impute (Rubin 1987) nonrespondent values of Y, with each set of imputations based on the propensity model applied to a bootstrap sample of the observations.

Like AIPW, PSPP has a DR property, in that it yields consistent estimates of the mean if either (A) the prediction model is correctly specified, or (B) both the propensity model is correctly specified <u>and</u> (C) the penalized spline correctly captures the relationship between the survey variable and the propensity; misspecification of the regression on other covariates does not yield bias, because of the balancing property of the propensity score noted in RR. This DR property has an additional condition (C) that is not required for consistency of AIPW, but arguably (C) is rendered plausible by the flexibility of the spline. The Bayesian version of PSPP thus has a form of DR without the need for weights, and it tends to have good confidence coverage under weak priors for the parameters. The method is conceptually simple, and the lack of weights avoid the messy practical issues of how to deal with extreme weights that lead to noisy estimates.

The methods in this and the previous subsection can also be used to handle selection bias, by replacing the indicator for nonresponse by the indicator for selection. See also Example 13.5 in Chapter 13. Selection bias and unit nonresponse can be handled simultaneously by applying the methods with the indicator for response R replaced by the indicator for selection and response.

15.4 Penalized Spline for Treatment Comparisons – PENCOMP

I now return to the original problem in RR, that is, controlling for confounders in an observational study comparing treatments. PENCOMP (Zhou, Elliott & Little 2019) applies the same basic model as PSPP to adjust for bias in estimated treatment effects due to observed confounders. The Neyman/Rubin causal model with T treatments defines the T outcomes under each treatment, with the $T - 1$ outcomes corresponding to the treatments not assigned being viewed as missing data. The probability of assignment to a treatment is estimated by a logistic or probit regression, or some other method appropriate for a binary outcome. A model involving the spline of the estimated propensity to be allocated treatment t and other covariates is then applied to predict the outcomes under treatments not assigned, and hence to estimate causal effects. A distinct model is applied to predict the outcomes for each treatment. A convenient implementation is to multiply-impute the outcomes under treatments not assigned and then base inference on multiple imputation combining rules (Rubin 1987). For cross-sectional observational studies, this yields a robust approach to estimating treatment effects.

PENCOMP can also handle *confounding by indication*, when treatments are assigned at more than one time point, and outcomes from treatments at an intermediate time point are used to determine treatment allocations at that time point. Simple regression methods do not apply because the intermediate outcomes are both outcomes (of initial treatment assignment) and

Unit,i	X_1	Z_1	$X_2^{(0)}$	$X_2^{(1)}$	Z_2	$Y^{(00)}$	$Y^{(01)}$	$Y^{(10)}$	$Y^{(11)}$
1	▨	0	▨	?	0	▨	?	?	?
2	▨	0	▨	?	0	▨	?	?	?
...	▨	...	▨	?	...	▨	...	...	...
n_{00}	▨	0	▨	?	0	▨	?	?	?
$n_{00}+1$	▨	0	▨	?	1	?	▨	?	?
$n_{00}+2$	▨	0	▨	?	1	?	▨	?	?
...	▨	...	▨	?	...	...	▨	...	...
$n_0=n_{00}+n_{01}$	▨	0	▨	?	1	?	▨	?	?
n_0+1	▨	1	?	▨	0	?	?	▨	?
n_0+2	▨	1	?	▨	0	?	?	▨	?
...	▨	...	?	▨	...	...	...	▨	...
n_0+n_{10}	▨	1	?	▨	0	?	?	▨	?
$n_0+n_{10}+1$	▨	1	?	▨	1	?	?	?	▨
$n_0+n_{10}+2$	▨	1	?	▨	1	?	?	?	▨
...	▨	...	?	▨	...	...	...	...	▨
$n=n_0+n_{10}+n_{11}$	▨	1	?	▨	1	?	?	?	▨

FIGURE 15.1
Confounding by indication with one intermediate outcome X_1. Principal stratification defines values of the intermediate outcome $X_2^{(0)}$ and $X_2^{(1)}$ under both treatments.

confounders (of later treatment assignment.) The key idea is to apply the Neyman/Rubin causal model to define outcomes to alternative treatments for both intermediate and final outcomes (Frangakis & Rubin 2002). With two treatments and one intermediate outcome, there are two possible outcomes for the intermediate time point, the one actually observed corresponding to the treatment assigned, and four possible outcomes at the final time point, of which one is observed corresponding to the treatment combination actually assigned. Figure 15.1 illustrates the data with baseline variables X_1, initial treatment assignment Z_1, intermediate outcome X_2, second treatment assignment Z_2, and final outcome Y. The shaded columns represent observed data, and clearly, a lot of data are missing! But multiply-imputing the missing data and applying MI combining rules provides valid and robust inferences that compare favorably with weight-based alternatives in simulation studies (Zhou, Elliott & Little 2019).

15.5 Positivity and Propensity Imbalance across Treatments

A key assumption of propensity methods is positivity – individuals have to have a positive probability of being assigned to all of the treatments being compared. In observational studies, this is far from a minor assumption – for particular combinations of covariates, one or more of the compared treatments might not to be assigned at all. If the estimated propensity to be assigned a particular treatment is zero, clearly a weight cannot be computed. If the propensity model yields very low propensities, they receive very large weights that can lead to very inefficient weighted estimates.

PSPP or PENCOMP somewhat ameliorates the problem of variable weights for reasons discussed above but does not resolve them. If the distribution of propensities differs substantially across two treatment groups, then the penalized spline model can still be fitted, but predictions involve extrapolation of the model into regions where there are little data, leaving results that are vulnerable to model misspecification. Thus, tools to reduce disparities in the propensity score distributions are important.

The task of PENCOMP is to adjust for observed confounders, where a confounder is a pre-treatment variable that is related to both treatment assignment and the outcome. Thus, variables that are (A) related to the outcome but not treatment assignment, or (B) related to treatment assignment but not to the outcome, are not true confounders. Adjusting for type (A) variables in the outcome model does not reduce bias but can increase precision, by reducing the residual variance. Adjusting for type (B) variables does not reduce bias but can reduce precision, particularly if included in a weighting adjustment. If PENCOMP is the method of adjustment, then including type (B) variables in the propensity model increases disparities in the propensity score distribution between treatment groups, which is still an undesirable effect. Simulations in Zhou, Elliott and Little (2021) show substantial improvements in PENCOMP estimates if type (B) variables are removed from the propensity model, or down-weighted using shrinkage methods such as the LASSO (Tibshirani 1996).

Applying such methods after data collection is subject to criticisms of "data snooping"; see, for example, Rubin (2007). However, biasing treatment effects can be minimized by prespecifying methods in detail in a study protocol before data collection, as is done in randomized clinical trials, and assessing potential confounders using a model that does not include treatment indicators as covariates.

The usual estimand for causal comparison is the ATE, the average treatment effect for the entire population of interest. However, when the degree of overlap in the propensity score distributions between treatment groups is limited, it may make sense to restrict the inference to subpopulations that do not have extreme response propensities close to zero or one. For example, Gutman and Rubin (2013) proposed dropping units outside of the overlap region of estimated propensity scores between treatment groups. The precise nature of a restricted estimand is unclear when sample cases are excluded based on estimated propensities, which vary in repeated sampling. However, restricting the estimand by dropping cases with extreme weights seems more defensible here than in the case of nonresponse discussed in Section 2, because, as Imbens and Wooldridge (2009) argue, the focus on the ATE estimand can be unrealistic, and comparisons might usefully be restricted to subpopulations where all compared treatments have a reasonable chance of being assigned. Fogerty et al. (2016) discuss how to define an interpretable study population wherein inference can be conducted without extrapolating with respect to important variables, when there is a lack of covariate overlap between the treated and control groups.

These considerations become more significant in longitudinal studies involving treatment allocations at multiple time points. In the AIDS application described by Zhou, Elliott and Little (2019), data are available at 16 time points, and either of two possible treatments could be assigned at each time point. Thus, there are over 30,000 (2^{15}) possible treatment combinations, nearly all of which are not seen in the data! Providing simple and interpretable causal conclusions in such a setting requires careful thought and modeling.

15.6 Some Thought Questions on This Chapter

1. Review the pros and cons of propensity modeling versus classical regression modeling for the adjustment of confounders in treatment comparisons.

2. Without referring to the text, prove for yourself the key properties of propensity score in Section 15.2.1, as a balancing score and as a method for simultaneously controlling for a set of confounders. (The proofs are simple but elusive, in my view.)

3. Find an observational study in a substantive area of interest to you that uses a propensity score method to control for confounders. Which of the methods of adjustment in Section 15.2.2 is used?

4. Suppose one of the X variables, say X_1, is a binary effect-modifier of the relationship between a treatment W and an outcome Y. Propose how distinct treatment effects for the two values of X_1 can be estimated adjusting for the other observed confounders, using classical regression and propensity score methods.

5. Review the matching methods for adjusting for confounders in Greifer and Stuart (2021). How do these methods compare with alternative methods based on regression or weighting?

6. Consider the methods discussed in this chapter in the light of the framework for treatment comparisons in Rubin (1978), discussed in Chapter 14. Specifically, consider how propensities for selection, assignment, and nonresponse might be used to address the gaps in Figure 14.1.

Appendix: Twenty Style and Grammar Tips for Statistics Writing

A feature arguably shared by all the seminal papers in this book is that they are well written; indeed, I have excluded some seminal papers because of poor writing style. As the son of a tabloid journalist, I have an appreciation for clear and succinct writing and try to achieve that in my own work. Good communication skills are important for all statisticians, and reading well-written papers can help develop these skills. As one of several good books on the subject, I recommend the concise and readable text on scientific writing style by Williams and Colomb (2010).

I have read many drafts of theses and papers written by biostatistics and statistics students and others, and certain recurring issues of English style arise. Below are 20 suggestions, offered to assist future writers and make life easier on future readers. Style is to some extent subjective, so the suggestions below are certainly not "theorems" (though some, like using a spell-checker, are close.) Also, a frustrating characteristic of the English language is that there are always exceptions to any rule. Some of the problems discussed below are particularly prevalent amongst non-native English speakers, for whom the problem of grappling with the statistics is compounded by the need to find ways of expressing ideas in a strange language. They have my genuine sympathy. But whether a native speaker or not, developing a good style should be a high priority, since it is an important ingredient for success in academia, government, or industry. The best way to improve is for non-native speakers to speak English, and for all to read widely and practice writing at every opportunity.

1. Tell a story! You are not writing a novel, but nevertheless try to craft a compelling story out of the work that motivates readers to read the paper. What is the problem that work addresses? Can you provide a concrete example? How does the work solve that problem? The story will vary greatly depending on the context, but a reader who is only mildly curious about the topic will not read the paper unless there is a compelling storyline.

2. Draft early and often. If you have done some work, or have an idea, write it down in a memorandum – to someone else who is interested (like your adviser), or otherwise to yourself! Don't keep work in your head until the end and then try to write it all out in one go. There are many reasons for this: the drafting of memoranda is a way of practicing writing, and it improves ideas by forcing clarity. An early draft can be continuously refined in a word processor, allowing wrinkles to be ironed out. Memoranda can form the basis for longer articles and eventually a thesis, so you have something to start with when writing them. Writing a thesis is an arduous task and should be spread out as much as possible, rather than waiting until the end.

3. Less is more! Scientific writing should be clear and concise. When you write anything, go over it carefully and attempt to reduce the length by 20% without reducing content, simply by more concise wording, removing adjectives that do not add anything, and avoiding repetition. Strive to say things just once clearly, rather than several times unclearly. (Avoid "Hmm, that didn't come out quite right, let's write it again another way…!") Some repetition is useful in lectures and more leisurely expositions, but for top academic journals, space is at a premium and repetition should be avoided. The result of this will be both shorter and better style.

When you have edited the piece once, go over it again and try to shorten it by another 10%!

Here is an example:

> Original: "The repeated-measures design has been recognized to permit direct study of change over time within individuals, and thus considered to be a better approach than the cross-sectional study design in estimating effects of covariates." (36 words)

> Edited: "Unlike cross-sectional designs, repeated-measures designs permit direct study of individual-level changes over time, and effects of covariates on these changes." (23 words)

Another example:

> Original: "In this section we apply our proposed models to the analysis of the data with compliance as the outcome of interest. The main objective of this analysis is to investigate how treatment preference and assignment preference affect the compliance status." (40 words)

> Edited: "In this section we apply our proposed models to the data with compliance as the outcome. The objective is to investigate how treatment and assignment preferences affect compliance." (28 words)

4. Know your audience. Before starting to write, make sure that you have a very clear idea of who is your intended audience, what you believe they already know, and what you think they might not know. Students often overestimate what the audience knows about the topic and their initial level of interest. As a result, more information on setting and motivation is often needed.

5. Avoid a pompous wordy style. Much scientific writing is pompous, complex, and wordy, as if this makes the author seem clever and the ideas important. Wrong! People love clear and direct writing.

6. Avoid a chatty conversational style. On the other hand, scientific writing is not like a dinner-time conversation. Non-natives who took classes in conversational English sometimes bring too much of that style to their scientific writing. This should be relatively formal, avoiding phrases like "Now let's prove this theorem …," "the estimate worked pretty well," and "the anemia profile differed a lot for infants …."

7. Use active rather than passive verbs. A major feature of the pompous, bad style is to distort the sentence order and use passive rather than active verbs, apparently thinking that this somehow sounds more objective and scientific. Try to use an active verb and avoid starting a sentence with long adjectival clauses, as is common in legal documents. Thus

> "The author calls this a prior likelihood approach" (8 words), NOT "This approach is referred to by the author as a prior likelihood approach" (13 words)

> "We choose a prior for f(t) similar to that used by Wahba (1978)" (13 words) NOT

> "based on a similar idea in Wahba (1978), we choose the same prior for f(t)" (15 words)

The "I" vs. "we" question. I am happy with "I" for single-authored works, but some editors seem to prefer the royal "we." I am not sure there is a clear consensus on this. For grants, avoid I and we – third person is more accepted.

8. Choose specific active verbs over generic verbs like "to do." For example,

> "The random-effects distribution has not yet been *modeled* nonparametrically" NOT

> "Modeling the random-effects distribution nonparametrically has not yet been *done*."

9. Don't pack too many ideas in one sentence and put the most interesting part for the reader at the end (the "stress position"). It is generally wise not to have more than one important idea in a sentence. We tend to expect the "payoff" or more interesting idea to be at the end of the sentence; if it is at the beginning, the rest of the sentence tends to be a let-down and may bore the reader.

10. Avoid repeating the same word in a short space. This can usually be avoided by rephrasing.

11. Seek a smooth flow of ideas, rather than jumping between topics. Review what you have written and check that there is no jumping around between topics without good reason. If A1 and A2 are on the same or similar topics, and B1 and B2 are on the same or similar topics, then order as A1A2B1B2 rather than A1B1A2B2.

12. Don't switch tenses in the same paragraph. For example, not:

> "After a trend in time was determined, both time-varying and time-stationary covariates will be added to the model." Rather

> "After a trend in time was determined, both time-varying and time-stationary covariates were added to the model."

In general, I use the present tense to describe work rather than the past tense, though the past tense is not wrong.

13. Use a spell/grammar checker! Typos are sloppy and many can be avoided using computer technology. If you are using LaTex and don't have a good spell-checker, read the text into a mainline word processor like Microsoft Word, and use the spell-checker in that program to pick up spelling and grammar problems in the pure text parts. Then save it as a text file and read it back into LaTex. You still need to check for typos – a spell-checker is not perfect.

14. Think hard about the best choice of notation and make sure *all* notation has been carefully and consistently defined. Use a single consistent notation throughout – if covariates are x in section 1, they should still be x in section 5, not z. It takes work to get this right!

15. Use section and equation numbers liberally and refer to them in the text. Numbering is a precise way of guiding readers through the paper.

16. Match singular/plural for subjects and verbs:

"random effects are assumed normal," not "random effects is assumed normal."

"in this work" or "in these works" NOT "in these work."

17. Articles for singular/plural nouns. Singular nouns generally require a definite or indefinite article, "the" or "a"; plural nouns generally do not. (This drives my Asian friends crazy!)

"lack of an explicit form for" NOT "lack of explicit form for."

"lack of explicit forms for" NOT "lack of the explicit forms for."

"it is difficult to evaluate the posterior loglikelihood of …" NOT "it is difficult to evaluate posterior loglikelihood of …"

18. Get the prepositions right. English is tricky, it takes practice to get it right.

"distribution for X" not "distribution on X."

"denote by X the random variable" or "let X denote the random variable" NOT "denote X the random variable."

19. Don't begin a sentence with a number. For example, start a sentence "Eighty percent of mothers …" NOT "80% of mothers …"

20. Miscellaneous pet peeves

20a. "The interesting parameters" or better "the parameters of interest" NOT "the interested parameters." (Parameters are inanimate objects and do not have interests.)

20b. The word "data" is plural. (In Latin, but this may be a losing battle.)

20c. "criteria" is plural. The singular is "criterion."

20d. "which" starts a nonrestrictive (descriptive) clause, "that" starts a restrictive clause. Often "which" is used when "that" would be more precise. Hint for native English speakers: if "that" sounds ok it is probably right. For example,

"hence d is the maximum likelihood estimate, ~~that~~ which has the useful property of being asymptotically efficient."

"Of these two estimators, the one ~~which~~ that satisfies the order restriction is better."

20e. "Alternate" (meaning to recur repeatedly; every other) is often used when the correct work is "Alternative" (meaning available as another possibility).

20f. "well" vs. "good" In informal speech, Americans often use the adjective "good" when the correct usage is the adverb "well." Hence "this was a good procedure" is OK, but "the procedure performed well," NOT "the procedure performed good."

20g. The founder of modern statistics is R.A. Fisher, not R.A. Fischer.

20h. Collective nouns are singular, not plural. For example, "Future work" NOT "Future works." "In future research we plan to study ..." NOT "In future researches we plan to study."

20i. "its" (no apostrophe) for possessives, "it's" (with apostrophe) for contractions of "it is." For example, "Although maximum likelihood is optimal asymptotically, ~~its~~ it's not so good in small samples." "Maximizing the likelihood with respect to ~~it's~~ its parameters yields ...."

Punctuation is a whole other story; for an amusing and informative read see Truss (2006).

References

Afifi, A. & Azen, S. (1979). *Statistical Analysis: a Computer-Oriented Approach*, 2nd edition. Academic Press.

Altham, P. M. E. (1969). Exact Bayesian analysis of a 2 × 2 contingency table and Fisher's exact significance test. *J. Roy. Statist. Soc., Ser. B*, 3, 1, 261–269.

Amberson, J. B., McMahon, B. T. & Pinner, M. (1931). A clinical trial of sanocrysin in pulmonary tuberculosis. *Amer. Rev. Tuberculosis*, 24, 401–435.

Anderson, T. W. (1957). Maximum likelihood estimates for the multivariate normal distribution when some observations are missing. *J. Am. Statist. Assoc.*, 52, 200–203.

Andridge, R. H. & Little, R. J. (2011). Proxy pattern-mixture analysis for survey nonresponse. *J. Official Statist.*, 27, 2, 153–180.

Baiocchi, M. & Rodu, J. (2021). Reasoning using data: two old ways and one new. *Observational Studies*, 7, 1, 3–12.

Baker, R. J. & Nelder, J. A. 1983. GLIM - Generalized linear models statistical software. In *Encyclopedia of Statistical Sciences*, Johnson, N. L. and Kotz, S. (eds.). Wiley.

Banks, D. (2021). Leo Breiman: a retrospective. *Observational Studies*, 7, 1,

Barnard, G. A. (1945). A new test for 2 × 2 tables. *Nature*, 156, 177.

Barnard, G. A. (1974). Discussion of Professor Stone's paper. *J. Roy. Statist, Soc. B*, 36, 133.

Barnett, V. (1973). *Comparative Statistical Inference*. London: Wiley.

Bartlett, M. S. (1936). The information available in small samples. *Proc. Cambridge Philosophic Soc.* 32, 560–566.

Basu, D. (1971). An essay on the logical foundations of survey sampling, Part 1. *Foundations of Statistical Inference*, 203–242. Holt, Rinehart & Winston: Toronto.

Bayarri, M. J., Benjamin, D., Berger, J. & Sellke, T. (2016). Rejection odds and rejection ratios: a proposal for statistical practice in testing hypotheses. *J. Math. Psych.*, 72, 90–103.

Bayarri, M. J. & Berger, J. O. (2000). P values for composite null models (with discussion). *J. Amer. Statist. Assoc.*, 95, 1127–1172, 13–15.

Behrens, W. V. (1929). Ein Betrag zur Fehlenbemhnung bei wenigen Beobachtungen. Landwirtschaftlich Jb. 68, 807–837.

Belson, W. A. (1959). Matching and prediction on the principle of biological classification. *Appl. Statist.*, 8, 65–75.

Benjamin, D. J. et al. (2018). Redefine statistical significance. *Nat. Hum. Behav.*, 2, 1, 6–10.

Benjamini, Y., De Veaux, R., Efron, B., Evans, S., Glickman, M., Graubard, B. I., He, X., Meng, X.-L., Reid, N., Stigler, S. M., Vardeman, S. B., Wikle, C. K., Wright, T., Young, L. J., & Kafadar, K. (2021). ASA President's Task Force statement on statistical significance and replicability. *Harv Data Sci Rev*, 3, 3. https://doi.org/10.1162/99608f92.f0ad0287

Benjamini, Y. & Hochberg, Y. (1995). Controlling the false discovery rate: a practical and powerful approach to multiple testing. *J. Roy. Statist. Soc. Ser. B*, 57, 1, 289–300.

Bennett, J., Lanning, S. et. al. (2007). The Netflix prize. In *Proc. KDD cup and workshop*, p35.

Berger, J. M. (2000). Bayesian analysis: a look at today and thoughts for tomorrow. *J. Amer. Statist. Assoc.*, 95, 1269–1276.

Berger, J. O. & Wolpert, R. L. (1988). *The Likelihood Principle*. Institute of Mathematical Statistics Lecture Notes-Monograph Series, 6, 1–199. Hayward, CA: Institute of Mathematical Statistics

Berkson, J. (1978) In dispraise of the exact test. *J. Statist. Planning & Inference*, 2, 27–42.

Bernado, J. M. (1979), "Reference posterior distributions for Bayesian inference" (with discussion). *J. Roy. Statist. Soc. Ser. B*, 41, 113–147.

Berry, D. (2012). Multiplicities in cancer research: ubiquities and necessary evils. *J. Nat. Cancer Inst.*, 104, 1124–1132.

Bickel, P. (2021). Comments on Breiman: statistical modelling: the two cultures and commentaries. *Observational Studies*, 7, 1, 17–20.

Birnbaum, A. (1962). On the foundations of statistical inference (with discussion). *J. Amer. Statist. Assoc.*, 57, 269–326.

Box, G. E. P. (1980). Sampling and Bayes' inference in scientific modelling and robustness. *J. Roy. Statist. Soc. Ser. A*, 143, 4, 383–430.

Breiman, L. (2001). Statistical modeling: two cultures. *Statist. Sci.* 16, 3, 199–231.

Box, G. E. P. & Tiao, G. C. (1973). *Bayesian Inference in Statistical Analysis*, 1st ed. New York: Addison-Wesley.

Breslow, N. E. & Clayton, D. G. (1993). Approximate inference in generalized linear mixed models. *J. Amer. Statist. Assoc.*, 88 (421), 9–25.

Brewer, K. (2013). Three controversies in the history of survey sampling. *Survey Meth.*, 39, 2, 249–262.

Brown, L. D., Cai, T. & DasGupta, A. (2001). Interval estimation for a Binomial proportion. *Statist. Sci.*, 16, 2, 101–133.

Buehler, R. J. (1959). Some validity criteria for statistical inference. *Ann. Math. Statist.* 30, 845–863.

Buja, A., Berk, R., Brown, L., George, E., Pitkin, E., Zhan, L. & Zhang, K. (2019). Models as approximations 1: consequences illustrated with linear regression. *Statist. Sci.*, 34, 4, 580–583.

Cameron, E. & Pauling, L. (1976). Supplemental ascorbate in the supportive treatment of cancer: prolongation of survival times in terminal human cancer. *Proc. Natl. Acad. Sci.*, 73, 10, 3685–3689.

Carter, G. M. & Rolph, J. E. (1974). Empirical Bayes methods applied to estimating fire alarm probabilities, *J. Amer. Statist. Assoc.*, 69, 880–885.

Carvalho, C. M., Polson, N. G. & Scott, J. G. (2010). The horseshoe estimator for sparse signals. *Biometrika*, 97, 2, 465–480.

Chaibub Neto, E., Bare, J. C. & Margolin, A. A. (2014). Simulation studies as designed experiments: the comparison of penalized regression models in the "large p, small n" setting. PLoS ONE 9(10), e107957.

Chalmers, I. (2011). Why the 1948 MRC trial of streptomycin used treatment allocation based on random numbers. *J. Roy. Soc. Med.*, 104, 383–386.

Chen, Q., Elliott, M.R., Haziza, D., Yuan, Y., Ghosh, M., Little, R. J., Sedransk, J. & Thompson, M. (2017). Approaches to improving survey-weighted estimates. *Statist. Sci.*, 32, 2, 227–248.

Chipman, H. A., George, E. I. & McCulloch, R. E. (2010). BART: Bayesian additive regression trees. *Ann. Appl. Statist.*, 4, 1, 266–298.

Cleland, J. G., Little, R. J. & Pitaktepsombati, P. (1978). Socioeconomic determinants of contraceptive use in Thailand. *WFS Scientific Reports, No. 5,* International Statistical Institute, Voorburg, The Netherlands.

Cook, R. J. & Farewell, V. T. (1996). Multiplicity considerations in the design and analysis of clinical trials. *J. Roy. Statist. Soc., Ser. A,* 159, 93–110.

Cox, D. R. (1958). Some problems connected with statistical inference. *Ann. Math. Statist.* 29 357–372.

Cox, D. R. (1965). A remark on multiple comparison methods. *Technometrics,* 7, 2, 223–224.

Cox, D. R. (1971). The choice between alternative ancillary statistics. *J. Roy. Statist. Soc. Ser. B,* 33, 251–255.

Cox, D. R. (1977). The role of significance tests. *Scand. J. Statist.,* 4, 49–70.

Cox, D. R. & Hinkley, D. V. (1974). *Theoretical Statistics.* Chapman & Hall, London.

Creagan, E. T et al. (1979). Failure of high-dose vitamin C (ascorbic acid) therapy to benefit patients with advanced cancer. *New Engl. J. Med.* 301, 13, 687–690.

Cruz-Cortés, E., Yang, F., Juaréz-Colunga, E., Warsavage, T. & Ghosh, D. (2021). Comment on 'Statistical modelling, the two cultures' by Leo Breiman. *Observational Studies,* 7, 1, 41–57.

David, M., Little, R. J., Samuhel, M. E. & Triest, R. K. Imputation models based on the propensity to respond. *Proc. Bus. Econ. Section, American Statistical Association 1983,* 168–173.

Davison, A. C. & Hinkley, D. V. (1997). *Bootstrap Methods and Their Application.* Cambridge University Press.

Dawid, A. P. (1982). The well-calibrated Bayesian. *J. Am. Statist. Assoc.,* 77, 605–610.

Dawid, A. P. (1984). Discussion of "On the Birnbaum argument for the Strong Likelihood Principle." *Statist. Sci.,* 29, 2, 240–241.

Dawid, A. P., Stone, M. & Zidek, J. V. (1973). Marginalization paradoxes in Bayesian and structural inference (with discussion)., *J. Roy. Statist. Soc. Ser. B,* 35, 189–233.

De Finetti, B. (1974). *Theory of Probability.* New York: John Wiley.

DeMets, D. L. & Ware, J. H. (1980). Group sequential methods for clinical trials with a one-sided hypothesis. *Biometrika,* 67, 651–60.

Dempster, A. P. (1980). Comment. *J. Amer. Statist. Assoc.,* 75, 372, 817–817.

Dempster, A. P., Laird, N. M. & Rubin, D. B. (1977). Maximum likelihood from incomplete data via the EM algorithm (with discussion), *J. Roy. Statist. Soc. B,* 39, 1–38.

Dempster, A. P., Schatzoff, M. & Wermuth, N. (1977). A simulation study of alternatives to ordinary least squares (with discussion). *J. Amer. Statist. Assoc.,* 72, 357, 77–106.

DiCiccio, T. J. & Efron, B. (1996). Bootstrap confidence intervals. *Statist. Sci.,* 11, 3, 189–228.

Donoho, D. (2017). 50 years of data science. *J. Comp. Graphical Statist.,* 26, 4, 745–766.

Draper, D. (1995). Assessment and propagation of model uncertainty. *J. Roy. Statist. Soc. Ser. B,* 57, 1, 45–70.

Durbin, J. (1970). On Birnbaum's theorem on the relation between sufficiency, conditionality & likelihood. *J. Amer. Statist. Assoc.* 65, 395–398.

Efron, B. (1979). Bootstrap methods: another look at the jackknife. *Ann. Statist.,* 7, 1, 1–26.

Efron, B. (2005), "Bayesians, frequentists and scientists," *J. Amer. Statist. Assoc.,* 100, 1–5.

Efron, B. & Morris, C. (1973). Stein's estimation rule and its competitors—an empirical Bayes approach. *J. Amer. Statist. Assoc.*, 68, 341, 117–130.

Efron, B. & Morris, C. (1977). Stein's paradox in statistics. *Scientific American*, 236, 5, 119–127.

Elliott, M. R. (2007). Bayesian weight trimming for generalized linear regression models. *Survey Meth.*, 33, 1, 23–34.

Elliott, M. R. & Little, R. J. (2000). Model-based approaches to weight trimming. *J. Official Statist.*, 16, 191–210.

Evans, M. J., Fraser, D. A. S. & Monette, G. (1986). On principles & arguments to likelihood. *Canad. J. Statist.* 14, 181–199.

Ferguson, T. S. (1967). *Mathematical Statistics: a Decision-Theoretic Viewpoint.* New York: John Wiley.

Firth, D. & Bennett, K. E. (1998). Robust models in probability sampling. *J. Roy. Statist. Soc., Ser. B,* 60, 3–21

Fisher, R. A. (1922a). On the mathematical foundations of theoretical statistics (with discussion). *Phil. Trans. Roy. Soc. London. Ser. A,* 222, 309–368.

Fisher, R. A. (1922b). On the interpretation of x2 from contingency tables, and the calculation of P. *J. Roy. Statist. Soc.,* 85, 87–94.

Fisher, R. A. (1935a). *The Design of Experiments.* Edinburgh: Oliver & Boyd.

Fisher, R. A. (1935b). The fiducial argument in statistical inference. *Ann. Eugenics,* 8, 391–398.

Fisher, R. A. (1945). The logical inversion of the notion of the random variable. *Sankhya* 7, 129–132.

Fisher, R. A. (1955). Statistical methods and scientific induction. *J. Roy. Statist. Soc., Ser. B,* 17, 1, 69–78.

Fogerty, C. B., Mikkelsen, M. E., Gaieski, D. F. & Small, D. S. (2016). Discrete optimization for interpretable study populations and randomization inference in an observational study of severe sepsis mortality. *J. Am. Statist. Assoc.,* 111, 514, 447–458.

Frangakis, C. E. & Rubin, D. B. (2002). Principal stratification in causal inference. *Biometrics,* 58, 1, 21–29.

Gelfand, A. E. & Smith, A. F. M. (1990). Sampling-based approaches to calculating marginal densities. *J. Amer. Statist. Assoc.,* 85, 410, 398–409.

Gelman, A. (2003). A Bayesian formulation of exploratory data analysis and goodness-of-fit testing. *Int. Statist. Rev.,* 71, 2, 369–382.

Gelman, A. (2007). Struggles with survey weighting and regression modeling (with discussion.) *Statist. Sci.,* 22, 2, 153–188.

Gelman, A. (2013). Two simple examples for understanding posterior p-values whose distributions are far from uniform. *Electronic J. Statist.,* 7, 2595–2602.

Gelman, A. (2021). Reflections on Breiman's two cultures of statistical modeling. *Observational Studies,* 7, 1, 95–98.

Gelman, A., Carlin, J. B., Stern, H. S., Dunson, D. B., Vehtari, A. & Rubin, D. B. (2013). *Bayesian Data Analysis,* 3rd edition. Chapman & Hall/CRC Press.

Gelman, A. & Hill, J. (2006). *Data Analysis Using Regression & Multilevel/Hierarchical Models.* Cambridge University Press.

Gelman, A., Hill, J. & Yajima, M. (2012). Why we (usually) don't have to worry about multiple comparisons. *J. Res. Educ. Effectiveness,* 5, 189–211.

Gelman, A., Jakulin, A., Pittau, M. G. & Su, Y.-S. (2008). A weakly informative default prior distribution for logistic & other regression models. *Ann. Appl. Stat.* 2, 4, 1360–1383.

Gelman, A. & Little, T. C. (1997). Poststratification into many categories using hierarchical logistic regression. *Survey Methodology*, 23, 127–135.

Gelman, A., Meng, X.-L. & Stern, H. (1996). Posterior predictive assessment of model fitness via realized discrepancies (with discussion). *Statist. Sinica*, 6, 733–807.

Gelman, A. & Rubin, D.B. (1992). Inference from iterative simulation using multiple sequences. *Statist. Sci.* 7, 4, 457–472.

Gelman, A. & Shalizi, C.R. (2013). Philosophy and the practice of Bayesian statistics. *Brit. J. Math. Statist. Psych.*, 66, 8–38.

Geman, S. & Geman, D. (1984). Stochastic relaxation, Gibbs distributions, and the Bayesian restoration of images. *IEEE Trans. Pattern Anal. & Machine Intelligence*, 6, 6, 721–741.

George, E. I. & McCulloch, R. E. (1997). Approaches for Bayesian variable selection. *Statist. Sinica*, 7, 339–373.

George, E. I., Ročková, V., Rosenbaum, P. R., Satopää, V. A. & Silber, J. H. (2017). Mortality rate estimation and standardization for public reporting: Medicare's Hospital Compare. *J. Amer. Statist. Assoc.*, 112, 519, 933–947.

Ghosal, S. & van der Vaart, A. (2017). *Fundamentals of Nonparametric Bayesian Inference*, 1st edition. Cambridge Series in Statistical and Probabilistic Mathematics, Cambridge University Press.

Ghosh, M. & Kim, Y.-H. (2001). The Behrens-Fisher problem revisited: a Bayes-frequentist synthesis. *Can. J. Statist.*, 29, 1, 5–17.

Ghosh, M., Reid, N. & Fraser, D. A.S. (2010). Ancillary statistics: a review. *Statist. Sinica*, 20, 1309–1332.

Gibbons, J. D. & Pratt, J. W. (1975). P-values: interpretation and methodology. *Amer. Statist.* 29, 20–25.

Greifer, N. & Stuart, E.A. (2021). Matching methods for confounder adjustment: an addition to the epidemiologist's toolbox. *Epidemiol. Rev.*, 43, 1, 118–129.

Gutman, R. & Rubin, D.B. (2013). Robust estimation of causal effects of binary treatments in unconfounded studies with dichotomous outcomes. *Statist. Med.*, 32, 1795–1814.

Hacking, A. (1965). *The Logic of Statistical Inference*. Cambridge University Press.

Hadfield, J. D. (2010). MCMC Methods for multi-response generalized linear mixed models. *J. Statist. Software*, 33, 2. See http://www.jstatsoft.org/

Hájek, J. (1971). Comment on a paper by D. Basu. *Foundations of Statistical Inference*. p236. Holt, Rinehart & Winston: Toronto.

Hansen, M. H., Madow, W. G. & Tepping, B. J. (1983). An evaluation of model-dependent and probability-sampling inferences in sample surveys (with discussion). *J. Amer. Statist. Assoc.*, 78, 776–793.

Hartley, H. O. & Rao, J. N. K. (1967). Maximum-likelihood estimation for the mixed analysis of variance Model. *Biometrika*, 54, 93–108.

Heckman, J. (1976). The common structure of statistical models of truncation, sample selection and limited dependent variables, and a simple estimator for such models. *Ann. Econ. Social Measurement*, 5, 475–492.

Hitchcock, D. B. (2009). Yates and contingency tables: 75 years later. *Electronic J. History Prob. Statist.*, 5, 2, 1–14.

Horvitz, D. G., & Thompson, D. J. (1952). A generalization of sampling without replacement from a finite universe. *J. Amer. Statist. Assoc.*, 47, 663–685.

Howard, J. V. (1998). The 2 × 2 table: A discussion from a Bayesian viewpoint. *Statist. Sci.* 13, 4, 351–367.

Hubbard, A. E., Ahern, J., Fleischer, N. L., Van Der Laan, M., Lippman, S. A., Jewell, N., Bruckner, T. & Satariano, W. A. (2010). To GEE or not to GEE. Comparing population average and mixed models for estimating the associations between neighborhood risk factors and health. *Epidemiology*, 21, 467–474.

Imbens, G. W. & Wooldridge, J. M. (2009). Recent developments in the econometrics of program evaluation. *J. Econ. Lit.*, 47, 5–86.

Isaki, C. T. & Fuller, W. A. (1982). Survey design under the regression superpopulation model. *.J. Amer. Statist. Assoc.*, 77, 89–96.

James, W. & Stein, C. (1961). Estimation with quadratic loss. *Proc. Fourth Berkeley Symp. Math. Statist. Prob.*, 1, 361–379.

Jennrich, R. I., & Schluchter, M. D. (1986). Incomplete repeated-measures models with structured covariance matrices, *Biometrics*, 42, 805–820.

Jensen, S. T., McShane, B. B. & Wyner, A. J. (2009). Hierarchical Bayesian modeling of hitting performance in baseball. *Bayesian Anal.*, 4, 4, 631–652.

Johnson, V. E. (2013). Revised standards for statistical evidence. *Proc. Natl Acad. Sci.* 110, 48, 19313–19317.

Kalbfleisch, J. (1975). Sufficiency and conditionality. *Biometrika*, 62, 251–268.

Kendall, M. (1959). Hiawatha designs an experiment. *Am. Stat.*, 13, 23–24.

Kish, L. (1995). The hundred years' wars of survey sampling. *Statistics in Transition*, 2, 813–830. Reproduced as Chapter 1 of *Leslie Kish: Selected Papers*, G. Kalton and S. Heeringa, (2003, eds.), New York: Wiley.

Kish, L. & Frankel, M. R. (1974). Inferences from complex samples (with discussion). *J. Roy. Statist. Soc., Ser. B*, 36, 1–37.

Koch, G. C., Landis, J. R., Freeman, J. L., Freeman, D. H. & Lehman, R. G. (1977). A general methodology for the analysis of repeated measurements of categorical data. *Biometrics*, 33, 133–58.

Laird, N. M. & Louis, T. A. (1989). Empirical Bayes ranking methods. *J. Educ. Statist.*, 14, 29–46.

Laird, N. M., & Ware, J. H. (1982). Random-effects models for longitudinal data, *Biometrics*, 38, 963–974.

Lazzeroni, L. C. & Little, R. J. (1998). Random-effects models for smoothing post-stratification weights. *J. Official Statist.*, 14, 61–78.

Leamer, E. (1978). *Specification Searches: Ad-Hoc Inferences with Experimental Data*. New York: Wiley.

Lehmann, E. L. (1993). The Fisher, Neyman-Pearson theories of testing hypotheses: one theory or two? *J. Amer. Statist. Assoc.* 88, 201–208.

Liang, K.-Y. & Zeger, S. L. (1986). Longitudinal data analysis using generalized linear models. *Biometrika*, 73, 1, 13–22.

Lightbourne, R., Singh, S. & Green, C. P. (1982). The World Fertility Survey: charting global childbearing. *Popul. Bull.*, 37, 1, 1–55.

Lindley, D. (1972). *Bayesian Statistics: A Review*. Philadelphia: SIAM.

Little, R. J. (1979). Maximum likelihood inference for multiple regression with missing values: a simulation study, *J. Roy. Statist. Soc. Ser. B*, 41, 76–87.

Little, R. J. (1986). Survey nonresponse adjustments. *Int. Statist. Rev.*, 54, 139–157.

Little, R. J. (1988). Some statistical issues at the World Fertility Survey. *The American Statistician*, 42, 31–36.

Little, R. J. (1989). On testing the equality of two independent binomial proportions. *Amer. Statist.*, 43, 283–288.

Little, R. J. (2004). To model or not to model? competing modes of inference for finite population sampling. *J. Amer. Statist. Assoc.*, 99, 546–556.

Little, R. J. (2006). Calibrated Bayes: a Bayes/frequentist roadmap. *Amer. Statist.*, 60, 3, 213–223.

Little, R. J. (2012). Calibrated Bayes: an alternative inferential paradigm for official statistics (with discussion and rejoinder). *J. Official Statist.*, 28, 3, 309–372.

Little, R. J. (2013). In praise of simplicity, not mathematistry! Simple, powerful ideas for the applied statistician. *J. Amer. Statist. Assoc.*, 108, 359–370.

Little, R. J. (2014). Survey sampling: past controversies, current orthodoxies, and future paradigms. In *Past, Present and Future of Statistical Science*, Lin, X., Banks, D. L., Genest, C., Molenberghs, G., Scott, D. W. and Wang, J.-L. (eds.). CRC Press.

Little, R. J. (2022). Bayes, buttressed by design-based ideas, is the best overarching paradigm for sample survey inference. *Survey Methodol.*, 48, 2, 257–281.

Little, R. J. (2022). Some reflections on Rosenbaum and Rubin's propensity score paper. *Observ. Stud.*, 9, 1, 69–75.

Little, R. J. & An, H. (2004). Robust likelihood-based analysis of multivariate data with missing values. *Statist. Sinica*, 14, 949–968.

Little, R. J. & Lewis, R. J. (2021). Estimands, estimators and estimates. *J. Amer. Med. Assoc.*, 326, 10, 967–968.

Little, R. J. & Rubin, D. B. (2019). *Statistical Analysis with Missing Data*, 1st edition. Wiley: New York.

Little, R. J. & Vartivarian, S. (2005). Does weighting for nonresponse increase the variance of survey means? *Survey Methodol.*, 31, 161–168.

Lyderson, S., Fagerland, M. W. & Laake, P. (2009). Tutorial in biostatistics: recommended tests for association in contingency tables. *Statist. Med.*, 28, 1159–1175.

Mayo, D. G. (2014). On the Birnbaum argument for the Strong Likelihood Principle (with discussion). *Statist. Sci.*, 29, 2, 227–266.

McCullagh, P. & Nelder, J. (1989). *Generalized Linear Models*, 2nd edition. Chapman & Hall.

McShane, B. B. & Gal, D. (2017). Statistical significance and the dichotomization of evidence, *J. Amer. Statist. Assoc.*, 112, 519, 885–895.

McShane, B. B, Gal, D., Gelman, A., Robert, C. & Tackett, J.L. (2019). Abandon statistical significance, *Amer. Statistician*, 73: suppl., 235–245.

McShane, L. M et al. (2005). Reporting recommendations for tumor marker prognostic studies (REMARK). *J. Natl. Cancer Inst.*, 97, 16, 1180–1184.

Medical Research Council (1948). Streptomycin treatment of pulmonary tuberculosis: a Medical Research Council investigation. *Brit. Med. J.*, 2, 769–782.

Mitra, N. (2021). Introduction to Special Issue: Commentaries on Breiman's Two Cultures paper. *Observational Studies*, 7, 1, 1–2, and the other papers in that volume.

Morgan, J. A. & Sonquist, J. N. (1963). Problems in the analysis of survey data: and a proposal. *J. Amer. Statist. Assoc.*, 58, 415–434.

Morral, A. R., Gore, K. L. & Schell, T. L. (2014). *Sexual Assault and Sexual Harassment in the U.S. Military: Volume 1. Design of the 2014 RAND Military Workplace Study.* Santa Monica, CA: RAND Corporation.

Moseley, J. B. et al. (2002). A controlled trial of arthroscopic surgery for osteoarthritis of the knee. *N. Eng. J. Med.* 347, 2, 81–88.

Nelder, J. (1975). Announcement by the Working Party on Statistical Computing: GLIM (Generalized Linear Interactive Modelling Program). *J. Roy. Statist. Soc. Ser. C*, 24, 2, 259–261.

Nester, M. R. (1996). An applied statistician's creed. *J. Roy. Statist. Soc. Ser. C*, 45, 4, 401–410.

Neuhauser, D. & Diaz, M. (2004). Shuffle the deck, flip the coin; randomization comes to medicine. *Quality & Safety Health Care*, 13, 315–316.

Neyman, J. (1934). On the two different aspects of the representative method: the method of stratified sampling and the method of purposive selection (with discussion). *J. Roy. Statist. Soc.*, 97, 4, 558–625.

Neyman, J. (1956). Note on an article by Sir Ronald Fisher. *J. Roy. Statist. Soc., Ser. B*, 18, 2, 288–294.

Neyman, J. & Pearson, E. S. (1933). On the problem of the most efficient tests of statistical hypotheses. *Phil. Trans. Roy. Soc. London, Ser. A*, 231, 289–337.

Ochi, Y. & Prentice, R. L. (1984). Likelihood inference in a correlated probit regression model. *Biometrika*, 71, 3, 531–543.

Park, T. & Casella, G. (2008). The Bayesian lasso. *J. Amer. Statist. Assoc.*, 103, 482, 681–686.

Pearson, E. S. (1955). Statistical concepts in their relation to reality, *J. Roy. Statist. Soc. Ser. B*, 17, 204–207.

Peers, H. W. (1965). On confidence points and Bayesian probability points in the case of several parameters. *J. Roy. Statist. Soc. B*, 27, 9–16.

Pepe, M. S. & Anderson, G. L. (1994). A cautionary note on inference for marginal regression models with longitudinal data and general correlated response data. *Commun. Statist. – Simul. Comp.*, 23, 4, 939–951.

Polack, F. P. et al. (2020) for the C4591001 Clinical Trial Group. Safety and efficacy of the BNT162b2 mRNA Covid-19 vaccine. *New Engl. J. Med.* 383, 2603–2615.

Pratt, J. W. (1977). 'Decisions' as statistical evidence and Birnbaum's 'confidence concept'. *Synthese*, 36, 59–69.

Preisser, J. S., Lohman, K. K & Rathouz, P. J. (2002). Performance of weighted estimating equations for longitudinal binary data with drop-outs missing at random. *Statist. Med.* 21, 20, 2025–3054.

Quenouille, M.H. (1949). Approximate tests of correlation in time series. *J. Roy. Statist. Soc. Ser. B*, 11, 68–84.

Raghunathan, T., Lepkowski, J., VanHoewyk, M., & Solenberger, P. (2001). A multivariate technique for multiply imputing missing values using a sequence of regression models, *Surv. Methodol.* 27, 1, 85–95.

Robins, J. & Wasserman, L. (2000). Conditioning, likelihood & coherence: a review of some foundational concepts. *J. Amer. Statist. Assoc.*, 95, 452, 1340–1346.

Robins, J. M., van der Vaart, A., & Ventura, V. (2000). Asymptotic distribution of P values in composite null models (with discussion). *J. Amer. Statist. Assoc.*, 95, 1143–1172.

Robinson, G. K. (1975). Some counter-examples to the theory of confidence intervals. *Biometrika*, 62, 155–161.

Robinson, G. K. (1976). Properties of Students t and of the Behrens-Fisher Solution to the two means problem. *Ann. Statist.*, 4, 5, 963–971.

Robinson, G. K. (1982). Behrens-Fisher problem. In *Encyclopedia of Statistical Sciences*, Vol. 6: Multivariate Analysis to Plackett & Burman Designs, Johnson, N. L., Kotz, S. and Read, C. B. (eds.), Wiley, New York, 205–209.

Rosenbaum, P. R. & Rubin, D. B. (1983). The central role of the propensity score in observational studies for causal effects. *Biometrika*, 70(1), 41–55.

Rosenbaum, P. R. & Rubin, D. B. (1984). Sensitivity of Bayes inference with data-dependent stopping rules. *The American Statistician*, 38, 2, 106–109.

Rothman, K. J. (1990). No adjustments are needed for multiple comparisons. *Epidemiology*, 1, 43–46.

Royall, R. M. & Cumberland, W. G. (1981). The finite population linear regression estimator and estimator of its variance-an empirical study. *J. Amer. Statist. Assoc.*, 76, 924–930.

Royall, R. M. & Cumberland, W. G. (1985). Conditional coverage properties of finite population confidence intervals. *J. Amer. Statist. Assoc.*, 80, 355–359.

Rubin, D. B. (1974). Estimating causal effects of treatments in randomized and non-randomized trials. *J. Educ. Psych.*, 66, 5, 688–701.

Rubin, D. B. (1976). Inference and missing data. *Biometrika*, 63, 581–592.

Rubin, D. B. (1978). Bayesian inference for causal effects: the role of randomization. *Ann. Statist.*, 6, 1, 34–58.

Rubin, D.B. (1980). Using empirical Bayes techniques in the law school validity studies (with discussion). *J. Amer. Statist. Assoc.*, 75, 372, 801–81

Rubin, D. B. (1981). The Bayesian bootstrap. *Ann. Statist.* 9, 1, 130–134.

Rubin, D. B. (1984). Bayesianly justifiable and relevant frequency calculations for the applied statistician. *Ann. Statist.*, 12, 4, 1151–1172.

Rubin, D. B. (1987). *Multiple Imputation for Nonresponse in Surveys*. Wiley: New York.

Rubin, D. B. (2007). The design versus the analysis of observational studies for causal effects: parallels with the design of randomized trials. *Statist. Med.*, 26, 20–36.

Rudin, C. (2019). Stop explaining black box machine learning models for high stakes decisions and use interpretable models instead. *Nature Machine Intelligence*, 1, 206–215.

Rudin, C. & Rudin, J. (2019). Why are we using black box models in AI when we don't need to? A lesson from an explainable AI competition. *Harvard Data Sci. Rev.*, 1, 2, 1–9.

Ruppert, D., Wand, M. P & Carroll, R. J. (2003). *Semiparametric Regression*. Cambridge University Press.

Samaniego, F. J & Reneau, D. M. (1994). Towards a reconciliation of the Bayesian and frequentist approaches to point estimation. *J. Amer. Statist. Assoc.*, 89, 947–957.

Satterthwaite, F. E. (1946). An approximate distribution of estimates of variance components. *Biometrics Bulletin*, 2, 6, 110–114.

Savage, L. J. (1954). *The Foundations of Statistics*. New York: John Wiley.

Schenker, N. (1985). Qualms about bootstrap confidence intervals. *J. Amer. Statist. Assoc.*, 80, 390, 360–361.

Scott, E. L. (1980). Comment. Rubin's empirical Bayes computations are not useful for law school admissions. *J. Amer. Statist. Assoc.*, 75, 372, 821–823.

Seidenfeld, T. (1992). R. A. Fisher's Fiducial argument & Bayes' theorem. *Statist. Sci.*, 7, 3, 358–368.

Shen, W. & Louis, T. A. (1998). Triple-goal estimates in two-stage hierarchical models. *J Roy. Statist. Soc. Ser. B*, 60, 2, 455–471.

Si, Y, Trangucci, R., Gabry, J. S. & Gelman, A. (2020). Bayesian hierarchical weighting adjustment and survey inference. *Survey Methodol.*, 46, 2, 181–214.

Smith, T. M. F. (1976). The foundations of survey sampling: a review (with discussion). *J. Roy. Statist. Soc. Ser. A*, 139, 183–204.

Smith, T. M. F. (1994). Sample surveys 1975-1990; an age of reconciliation (with discussion)? *Int. Statist. Rev.*, 62, 5–34.

Sprague, N. F. (1981). Arthroscopic debridement for degenerative knee joint disease *Clin. Orthopaedics*, 160, 118–123.

Stein, C. (1956). Inadmissibility of the usual estimator for the mean of a multivariate distribution, *Proc. Third Berkeley Symp. Math. Statist. Prob.*, 1, 197–206.

Stigler, S. (2005). Fisher in 1921. *Statist. Sci.*, 20, 1, 32–49.

Stiratelli, R., Laird, N. & Ware, J. H. (1984). Random-effects models for serial observations with binary response. *Biometrics*, 40, 4, 961–971.

Storey, J. D. (2003). The positive false discovery rate: a Bayesian interpretation and the q-value. *Ann. Statist.*, 31, 6, 2013–2035

Stuart, A. (2010). *Kendall's Advanced Theory of Statistics, 3 Volume Set*, 6th edition, New York: Wiley.

Suissa, S. & Shuster, J. J. (1989). Exact unconditional sample sizes for the 2 × 2 binomial trial. *J. Roy. Statist. Soc. Ser. A.*, Vol. 148, 4, 317–327.

Szpiro, A. A., Rice, K. M. & Lumley, T. (2010). Model-robust regression and a Bayesian "sandwich" estimator. *Ann. Appl. Statist.*, 4, 4, 2099–2113.

Taleb, N. N. (2007). *The Black Swan: the Impact of the Highly Improbable*. Random House.

Tanner, M. A. & Wong, W. H. (1987). The calculation of posterior distributions by data augmentation. *J. Amer. Statist. Assoc.*, 82, 398, 528–540.

Tibshirani, R. (1996). Regression shrinkage and selection via the lasso. *J. Roy. Statist. Soc. Ser. B*, 58, 1, 267–288.

Truss, L. (2006). *Eats, Shoots and Leaves: the Zero Tolerance Approach to Punctuation.* Avery Press.

Tukey, J. W. (1962). The future of data analysis. *Ann. Math. Statist.*, 33, 1, 1–67.

Tukey, J. W. (1977). *Exploratory Data Analysis*. 1st edition. Pearson Press.

Tukey, J. W. (1991). The philosophy of multiple comparisons. *Statist. Sci.*, 6, 1, 100–116.

Turkman, M. A. A., Paulino, C. D. & Müller, P. (2019). *Computational Bayesian Statistics: An Introduction*, 1st edition. Cambridge University Press.

Valliant, R., Dorfman, A. H. & Royall, R. M. (2000). *Finite Population Sampling and Inference: a Prediction Approach*. New York: Wiley.

van Buuren, S. & Groothuis-Oudshoorn, K. (2011). MICE: multivariate imputation by chained equations in R. *J. Statist. Software*, 45, 3, 1–67.

van der Laan, M. J., Polley, E. C. & Hubbard, A. E. (2007). Super learner. *Statist. Appl. Genetics Mol. Biol.*, 6, 1 Article 25.

Vansteelandt, S. (2021). Statistical modelling in the age of data science. *Observational Studies*, 7, 1, 217–228.

Vartivarian, S. & Little, R. J. (2003). Weighting adjustments for unit nonresponse with multiple outcome variables. *Proc Survey Res Methods Section Am Stat Assoc. 2003*, 4358–4363.

Wakefield, A. J., et al. (1998). Ileal-lymphoid-nodular hyperplasia, non-specific colitis, and pervasive developmental disorder in children. *Lancet*, 351, 637–641. (Later retracted.)

Walicke, P. et al. (2017). Launching effectiveness research to guide practice in neurosurgery: a National Institute of Neurological Disorders & Stroke workshop report. *Neurosurgery*, 80:4, 505–514

Wasserstein, R. L. & Lazar, N. A. (2016). The ASA's statement on p-values: context, process, and purpose. *Amer. Statist.*, 70, 2, 129–133, with supplemental comments at: https://www.t&fonline.com/doi/full/10.1080/00031305.2016.1154108?cookieSet=1

Wasserstein, R. L., Schirm, A. L. & Lazar, N. A. (2019). Moving to a world beyond p < 0.05. *Amer. Statistician*. 73, suppl., 1–19.

Weiss, L. (1955). A note on confidence sets for random variables. *Ann. Math. Statist.*, 26, 1, 142–144.

Welch, B. L. (1938) The significance of the difference between two means when the population variances are unequal. *Biometrika*, 29, 350–62.

Welch, B. L. (1956). Note on some criticisms made by Sir Ronald Fisher. *J. Roy. Statist. Soc. Ser. B*, 18, 2, 297–302.

Welch, B. L. (1965). On comparisons between confidence point procedures in the case of a single parameter. *J. Roy. Statist. Soc. B*, 27, 1–8.

Wilkinson, G. N. (1977). On resolving the controversy in statistical inference (with discussion). *J. Roy. Statist. Soc. Ser. B*, 39, 119–171.

Williams, J. M. & Colomb, G. G. (2010). *Style: Lessons in Clarity and Grace*, 10th Edition. Pearson Press.

Yang, Y. & Little, R. J. (2015). A comparison of doubly robust estimators of the mean with missing data. *J. Statist. Comp. Sim.*, 85, 16, 3383–3403.

Yates, F. (1934). Contingency tables involving small numbers and the 2×2 test. *J. R. Statist. Soc. Suppl.*, 1, 217–235.

Yates, F. (1984). Tests of significance for 2×2 contingency tables (with discussion). *J. Roy. Statist. Soc. Ser. A*, 147, 426–463.

Zabell, S. L. (1992). R.A. Fisher and the Fiducial argument. *Statist. Sci.* 7, 3. 369–387.

Zanganeh, S. & Little, R. J. Bayesian inference for the finite population total in heteroscedastic probability proportional to size samples. *Journal of Survey Statistics and Methodology*, 3:162–192, 2015.

Zeger, S. L. & Liang, K.-Y. (1992). An overview of methods for the analysis of longitudinal data. *Statist. Med.*, 11, 1825–1839.

Zeger, S. L., Liang, K. Y. & Self, S. G. (1985). The analysis of binary longitudinal data with time independent covariates. *Biometrika*, 72, 31–8.

Zhang, G. & Little, R. J. (2009). Extensions of the penalized spline of propensity prediction method of imputation. *Biometrics*, 65, 3, 911–918.

Zhang, G. & Little, R. J. (2011). A comparative study of doubly-robust estimators of the mean with missing data. *J. Statist. Comp. Sim.*, 81, 12, 2039–2058.

Zheng, H. & Little, R. J. (2003). Penalized spline model-based estimation of the finite population total from probability-proportional-to-size samples. *J. Official Stat.*, 19(2), 99–117.

Zheng, H. & Little, R. J. (2005). Inference for the population total from probability-proportional-to-size samples based on predictions from a penalized spline nonparametric model. *J. Official Statist.*, 21, 1–20.

Zhou, T., Elliott, M. R. & Little, R. J. (2019). Penalized spline of propensity methods for treatment comparisons (with discussion). *J. Amer. Statist. Assoc.*, 114, 525, 1–38.

Zhou, T., Elliott, M. R. & Little, R. J. (2021). Robust causal estimation from observational studies using penalized spline of propensity score for treatment comparison. *Stats*, 4, 529–549.

Zou, H. & Hastie, T. (2005). Regularization and variable selection via the elastic net. *J. Roy. Statist. Soc. Ser. B*, 67, 301–320, with correction on p. 768.

Author Index

Subject Index

For Product Safety Concerns and Information please contact our EU
representative GPSR@taylorandfrancis.com
Taylor & Francis Verlag GmbH, Kaufingerstraße 24, 80331 München, Germany

www.ingramcontent.com/pod-product-compliance
Lightning Source LLC
Chambersburg PA
CBHW060404220326
41598CB00023B/3011